Ceramography

Preparation and Analysis of Ceramic Microstructures

Richard E. Chinn

JOHN WILEY & SONS, INC.

The American Ceramic Society

www.acers.org

ASM INTERNATIONAL
The Materials Information Society

www.asminternational.org

First printing, December 2002

Comments, criticisms, and suggestions are invited, and should be forwarded to ASM International.

ASM staff who worked on this project include Scott Henry, Assistant Director of Reference Publications; Bonnie Sanders, Manager of Production; and Nancy Hrivnak and Jill Kinson, Production Editors.

Library of Congress Cataloging-in-Publication Data

Chinn, Richard E.
Ceramography : preparation and analysis of ceramic microstructures / Richard E. Chinn.
p. cm
Includes bibliographical references and index.
1. ceramography. I. Title.

TP811.5 .C45 2002
620.1'404299—dc21

2002027843

ISBN 978-0-87170-770-3
SAN: 204-7586

ASM International®
Materials Park, OH 44073-0002
www.asminternational.org

To Patsy, Melissa, and Heidi

Contents

Preface

This monograph is an extension of a paper I wrote in 1998 for a contest in *Structure*, the Struers Journal of Materialography that is published in Denmark. Although I was employed as a process engineer at a reactive-metals producer at the time, the paper was based on a project that began in the late 1980s when I worked for Coors Ceramics Company and Coors Brewing Company in a corporate materials laboratory in Golden, Colorado. The Coors companies now have completely separate management and stockholders, but I spent over a decade preparing and examining microstructures for the company now known as CoorsTek, Inc. and its subsidiaries in the ceramics industry. That aforementioned project resulted in a paper that was published by the *Journal of the American Ceramic Society* and two binders full of notes, drawings, and micrographs. I used some of these notes and techniques that I had developed in my master's thesis. The *Journal* paper, the *Structure* paper, and parts of the thesis became the basis of this book.

This book is hardly an exhaustive research summary, but it does contain a list of references at the end of each chapter. The references include many of the more common metallography books and papers from the ceramic journals that detail novel etching techniques and measurement methods. The Suggested Reading lists direct the reader to more comprehensive sources of information, primarily engineering textbooks, on topics that are presented here only in brief.

In regard to the vernacular of ceramography, *optical* microscope is not used, because all microscopes, even electron microscopes, are optical instruments. *Microscope*, in this text, implies a visible-light microscope, and any other type of microscope is specified explicitly. All the micrographs in this book are bright-field images, except as noted. (*Bright-field* is the form of illumination that causes specularly reflected surfaces normal to the axis of the microscope to appear bright.) Most of the specimens are from actual ceramic devices manufactured in the late 1980s to the mid-1990s by numerous companies around the world. These devices were pulled from the production line for quality-control purposes or sent in from the field for failure analysis. The fabrication methods, where known, are provided in the figure captions. A few of the specimens are experimental compositions and are noted as such.

I thank the many colleagues and mentors who have aided in my understanding of ceramics and optical instruments over the years. Among them are Doris Cheever; the late Dan Briggs; John McLane; Terry Brog; Mike Readey and Dan Schmidt at Coors; Professors Mike Haun, Dennis

Readey, and Bob Frost; fellow graduate students Dave Price and Chan Young Kim at the Colorado School of Mines; Seaton McLennan at Linn-Benton Community College; and James Bennett, Keith Collins, Cheryl Dahlin, Paul Danielson, Cindy Doğan, Dale Govier, Al Hunt, Larissa Iverson, Jasper Kwong, Joe Perry, and Art Petty at the Albany Research Center.

I also thank Charles A. Parker of Honeywell Aircraft Landing Systems and Roy W. Rice for their critical reviews of the manuscript; Scott Henry, Bonnie Sanders, and their staff at ASM International for their attention to the publishing details; and the correspondents named in some of the figure captions who generously contributed images from their research.

Richard E. Chinn

About the Author

Richard Chinn earned his B.Sc. in metallurgical engineering and M.Sc. in materials science at the Colorado School of Mines in Golden, Colorado, USA. He worked for the Coors companies in Golden for a decade as a materials engineer, where he gained experience in both technical ceramics and aluminum beverage cans. He was a process engineer at a ferroalloys smelter in Ohio and a reactive-metals smelter in Oregon before joining the United States Department of Energy's Albany Research Center as a materials research engineer in 2000. He is a registered professional engineer in Colorado and Oregon and has been a certified quality engineer of the American Society for Quality since 2000. He also holds credentials from the American Society of Nondestructive Testing and the Materials Engineering Institute. He is active in the American Ceramic Society and the Oregon chapter of ASM International.

List of Subroutines

CHAPTER 1

Introduction

Ceramography in Materials Science

CERAMOGRAPHY IS the art and science of preparation, examination, and evaluation of ceramic microstructures. The *microstructure* is the structure level of approximately 0.1 to 100 μm between the wavelength of visible light and the resolution limit of the naked eye. The microstructure includes most grains, secondary phases, grain boundaries, pores, microcracks, and hardness microindentations. The *micrograph* on the cover shows some of these features. A number of mechanical, optical, thermal, electrical, and other bulk properties of ceramics are significantly affected by the microstructure. Research papers often attempt to correlate the properties and observed behavior of experimental materials with the characteristics of the microstructure. The fabrication method and process conditions are generally reflected in the microstructure. The root cause of many ceramic failures is evident in the microstructure.

The 100 μm level is approximately the limit of resolution by the unaided human eye. The structure level larger than 100 μm is sometimes called the *macrostructure,* and some ceramics have grains and pores that are visible to the naked eye. A microscope cannot resolve anything significantly smaller than the wavelength of visible light, 360 to 780 nm, or roughly 0.4 μm. The electron microscopes, which use a beam of electrons rather than visible light to generate an image, are well suited for observation of microstructural features that are smaller than the wavelength of visible light.

Ceramography is part of the broader field of *materialography,* which includes all the microscopic techniques of material analysis, such as metallography, petrography, and plastography. Several notable books, chapters, and papers have been written about metallography, including books by Vander Voort and Petzow and an entire *ASM Handbook* Volume. Much less has been written about ceramography, but it is just as important to ceramic technology as metallography is to the metallurgical industries. Fulrath and Pask edited two conference proceedings centered on ceramic microstructures, especially correlations between microstructures and processing and properties. Buehler Ltd. offers a seminar on ceramography that includes a comprehensive manual. Clinton's book provides an overview of the development of ceramographic preparation up through the mid-1980s. The book by Elssner et al., translated from German, has many similarities to this one but with a special emphasis on composites containing ceramics, plasma spray coatings, and European literary references.

The scope of this text is limited to methods and applications of ceramography, that is, microstructures of ceramics. The techniques for ceramographic preparation are divided into five parts: sawing, mounting, grinding, polishing, and etching. A number of microscopic techniques for qualitative and quantitative analysis are discussed, along with test methods, equipment manufacturers, and other information sources. Many other related topics are mentioned but not described in great detail, including crystallography, electron microscopy, hardness, and ceramic processing. References to research papers and textbooks are provided for further inquiry. A flow chart of ceramographic preparation and analysis is shown in Fig. 1.1.

Crystallography

Crystallography, the study of crystal structures and lattice spacing, is an important adjunct to ceramography in phase identification and research. Crystallography is studied primarily by x-ray diffraction and electron diffraction. Crystallography is beyond the scope of this book, although it is alluded to occasionally. Crystallography is used to identify crystal structures (Bravais lattice + basis) and compounds as opposed to individual elements. Crystallography does not provide images of molecules or atoms. The crystal structures of some common ceramics are given in Table 1.1.

A finer level of structure than the microstructure could be called the nanostructure, although that term has other connotations. This level begins at the diameter of an atom, on the order of 0.1 nm. Some types of electron and scanning probe microscopes, mentioned in the section "Other Microscope Types" in Chapter 7, can resolve dislocations, stacking faults, and even atoms. The structure levels between molecules and microstructure contain colloids, which are solid particles so fine that their surface energy becomes a key property. Ceramic powders containing 50 nm particles are used for specialized applications such as capacitors.

The subatomic structure is yet another level of structure. Each level has certain effects on the properties of materials. X-ray crystallography is described in detail in Suryanarayana and Norton's book. Ceramics at the molecular level and optical properties of ceramics are discussed in books by Callister; Smith; and Kingery, Bowen, and Ullmann. Crystal structures of additional ceramics are provided by Bonnell.

Laboratory Safety

Ceramographic preparation involves a number of mechanical, thermal, chemical, and electrical hazards. Most safety items are readily available from laboratory supply companies and industrial distributors. In general, the personal protective equipment and safe practices to be used include:

- Use safety glasses with side shields for all sawing, grinding, polishing, and etching procedures. These operations can generate airborne particles and splashed liquids.
- Read and thoroughly understand the material safety data sheet (MSDS) of all chemicals, solvents, resins, ceramic powders, abrasive slurries, and finished ceramics.

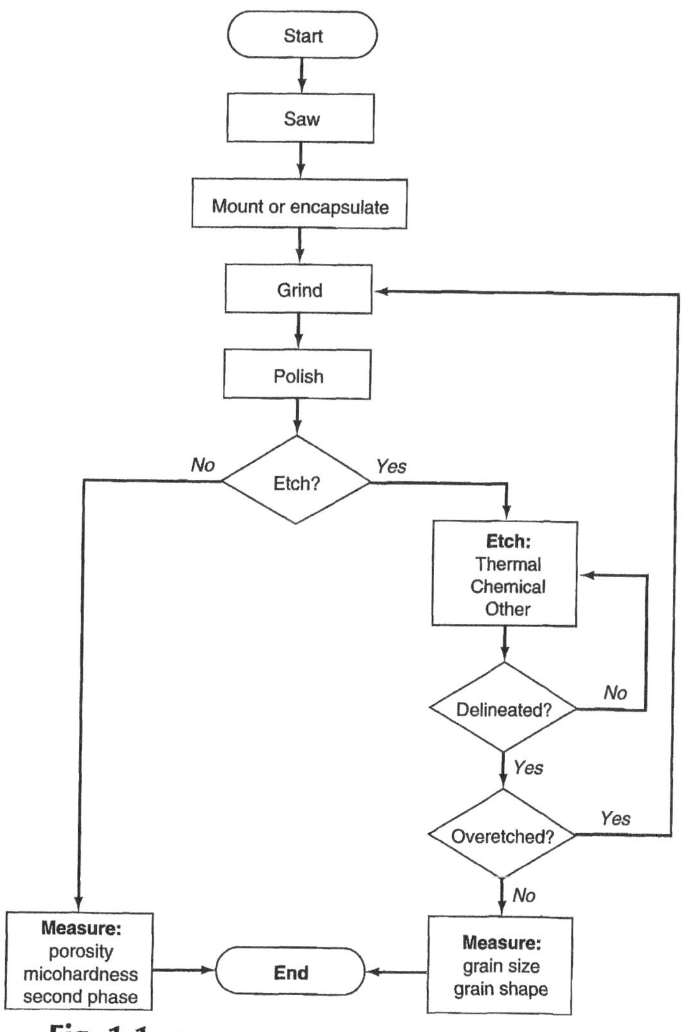

Fig. 1.1 Flow chart of ceramographic preparation and analysis

- Use a face shield over the safety glasses, especially for chemical etching. The face shield provides an additional layer of protection from splashes, sprays, and flying particles.
- Wear skin-tight latex or nitrile gloves for manual grinding and polishing. Both the water and the fine, abraded particles can be very irritating to bare skin after repeated exposure.
- Use thick rubber gloves, a rubber apron, and a fume hood for chemical etching. Many chemical etchants, especially those containing hydrofluoric acid ($HF_{(aq)}$), attack human tissue rapidly and can cause severe injury.
- Use fiberglass-insulated gloves and steel tongs for thermal etching, to avoid burns.
- Properly ground all electrical equipment. Many ceramographic machines employ both electricity and water and must be grounded in compliance with electrical codes.
- Properly store and dispense with all chemical reagents, resins, and solvents. Approved, labeled containers and cabinets must always be used to prevent explosions, spills, contamination, and other disasters. Food, beverages, medicine, and tobacco must not be consumed or stored in the laboratory at any time. Chemicals must never be used or stored in the lunchroom.
- Properly dispose of all chemical etchants, solvents, swarf, and other wastes. The United States Environmental Protection Agency and most state and county health departments do not permit these wastes in sewers, because they can contaminate groundwater and corrode and clog plumbing.

Table 1.1 Crystal structures of some common ceramics

Ceramic	Ceramic name	Mineral names	Crystal structure(a)	Space group
α-Al_2O_3	Alpha alumina	Corundum, sapphire	Hexagonal (corundum)	$R3c$
AlN	Aluminum nitride	...	Hexagonal (wurtzite)	$P6_3mc$
$Al_6Si_2O_{10}$	Mullite	Mullite	Orthorhombic	$Pbam$
B_4C	Boron carbide	...	Rhombohedral	$R\bar{3}m$
$BaTiO_3$	Barium titanate	...	Orthorhombic (perovskite)	$Pcmn$
C	Diamond	Diamond	Face-centered cubic (diamond)	$F43m$
C	Graphite	Graphite	Hexagonal	$P6_3/mmc$
$MgAl_2O_4$	Spinel	Spinel	Face-centered cubic (spinel)	$Fd3m$
MgO	Magnesia	Periclase	Cubic (rock salt)	$Fm\bar{3}m$
SiO_2	Silica	Quartz	Hexagonal	$P3_221$
α-SiC	Alpha silicon carbide	...	Hexagonal (wurtzite)	$P6_3mc$
β-SiC	Beta silicon carbide	...	Face-centered cubic (diamond) or (zinc blende)	$F43m$
β-Si_3N_4	Beta silicon nitride	...	Hexagonal	$P6_3/m$
TiB_2	Titanium diboride	...	Hexagonal	$P6/mmm$
ZrO_2	Zirconia, fully stabilized with Y_2O_3 in solid solution (ZDY or CSZ)	Baddeleyite (as monoclinic ZrO_2)	Face-centered cubic (metastable) (fluorite)	$Fm3m$
ZrO_2	Tetragonal zirconia polycrystals (TZP)	Baddeleyite (as monoclinic ZrO_2)	Tetragonal (metastable)	$P4_2/nmc$

Note: ZDY, zirconia densified with yttria; CSZ, cubic stabilized zirconia. (a) The structure names in parentheses are used to characterize the combined anion and cation Bravais lattices.

- Use a fume hood with the exhaust fan actuated for operations that generate toxic vapors, fumes, mist, or dust. Many of the chemical etchants and organic solvents are very volatile, especially when heated. The vapors may be reactive, poisonous, or flammable. Calcination of plastic encapsulations (see Subroutine 5.1) generates smoke that should be exhausted.
- Contain toxic or radioactive dust, even in small quantities. Most of the ceramographic sawing and abrasive operations use water or other liquids as lubricants, such that the dust that would be generated in a dry operation is contained. Nonetheless, beryllia (BeO) dust can cause a disabling and sometimes fatal respiratory disease called berylliosis. Silica (SiO_2) and silicate dust can cause silicosis. Radioactive ceramics, such as urania (UO_2), thoria (ThO_2), and plutonia (PuO_2), have been linked to cancer. Carcinogenic materials such as these should be processed only in a glove box or hot cell or as required by law.

Ceramographic Laboratory Design

In many factories and universities, the ceramographic laboratory is an afterthought, subject to whatever space is available. Wet-chemistry laboratories, on the other hand, are often well planned and equipped accordingly. The safety, efficiency, and utility of a ceramographic laboratory depend on a good layout of rooms and equipment.

At the very least, the optical instruments should be located in a separate room from the preparation equipment. The microscopes, image analyzers, and hardness testers should be kept in a room that can be closed and locked and is relatively free from vibrations and dust sources. The instrument room should be as far away as possible from the production machinery in a factory. A concrete floor is desirable although not absolutely necessary. The instruments require "clean" electricity and should be surge-protected and grounded. Instruments mounted on air suspension tables need a supply of compressed air and a regulator.

The electron microscopes are especially sensitive to external disturbances and should be located in their own air-conditioned room. A concrete floor, acoustically isolated from production machinery and motor vehicles, is recommended. Electron microscopes have lots of cables and hoses that must be located away from walkways. An adjacent utility room is desirable for the location of the vacuum pump and any other noisy accessories. X-ray spectrometers usually require liquid nitrogen on a daily basis, so the electron microscope room should be close to the liquid nitrogen source. The liquid nitrogen tank should be located where it can be easily exchanged for a full tank, such as near a door to a driveway. The sputter coater uses compressed argon and must be located near a compressed gas cylinder support.

Table 1.2 Utilities and accessories for the ceramographic laboratory

Utility or accessory	Usage
115 and 230 V_{ac} ac electricity	Grinding and polishing machinery, saws, hotplates, microscopes, mounting presses, microindentation hardness esters, image analyzers, sputter coaters, vacuum pumps, air compressors, furnaces, ultrasonic tanks, heat guns
460 V_{ac} 3 ϕ electricity	Some larger furnaces and motorized machinery
Cold water	Grinding machinery, saws, automatic mounting press, specimen washing sink, chemical etching
Hot water	Specimen washing sink and chemical etching
Distilled or deionized water	Specimen washing and chemical etching
Sewer drain	Grinding and polishing machinery, saws, automatic mounting press, specimen washing sink
Compressed air	Grinding and polishing machinery, automatic mounting press, air suspension table, some Bunsen burners, castable-resin mounting
Methane	Bunsen burners, some furnaces
Exhaust duct	Fume hood for chemical etching and encapsulation removal, storage cabinets for solvents and chemicals, gas-fired furnace
Vacuum pump	Electron microscopes, castable-resin mounting, some furnaces
Compressed gas cylinder support	Etching furnace, sputter coater
Liquid nitrogen	X-ray spectrometer for electron microscopes

ac, alternating current

The preparation machines require multiple utilities and must be located in a room where the utility connections are easily accessible or can be installed. The utilities vary from one machine to another but may include 115 and 230 V_{AC} electricity, 460 V_{AC} 3 ϕ electricity, cold water, hot water, compressed air, methane, sewer drains, exhaust ducts, and vacuum. A ground-floor room with a concrete floor and double-door or garage-door access to a driveway is ideal. A separate room for the air compressor and vacuum pump is desirable to minimize noise.

An etching room, separate from the preparation equipment, is desirable. Most etching furnaces require the higher-voltage electricity and need compressed gas cylinder supports nearby, if nitrogen or argon is to be used to purge the furnace. The etching chemicals and solvents should be stored in safety cabinets well away from heat sources, including the furnace. A fume hood with a working exhaust fan is absolutely necessary for chemical etching and encapsulation removal by burning. A linear or face flow rate of at least 0.5 m/s (100 ft/min) in the hood exhaust is recommended. The fume hood should also have a sink with hot and cold water and a methane port. The etching room should be located close to the restricted area where waste chemicals are stored, for easy access, and far from where food is stored and consumed.

Laboratory furniture is distributed by some of the equipment manufacturers listed in Appendix B and also by the laboratory supply firms. A thermometer, barometer, and relative humidity gage in the instrument room are an asset to any laboratory. The laboratory utilities and accessories are summarized in Table 1.2.

SUGGESTED READING

- D.A. Bonnell, Crystal Structures, *Ceramics and Glasses,* Vol 4, *Engineered Materials Handbook,* ASM International, 1991, p 879–882
- W.D. Callister, Jr., *Materials Science and Engineering: An Introduction,* 5th ed., John Wiley & Sons, Inc., 2000
- D.J. Clinton, *A Guide to Polishing and Etching of Technical and Engineering Ceramics,* The Institute of Ceramics, 1987
- G. Elssner, H. Hoven, G. Kiessler, and P. Wellner, *Ceramics and Ceramic Composites: Materialographic Preparation,* R. Wert, Trans., Elsevier Science Inc., 1999
- R.M. Fulrath and J.A. Pask, Ed., *Ceramic Microstructures: Their Analysis, Significance, and Production,* Robert E. Krieger Publishing Company, 1968
- R.M. Fulrath and J.A. Pask, Ed., *Ceramic Microstructures '76: With Emphasis on Energy Related Applications,* Westview Press, Inc., 1977
- C.A. Johnson, *Metallography Principles and Procedures,* Leco Corp., 1991
- W.D. Kingery, H.K. Bowen, and D.R. Uhlmann, *Introduction to Ceramics,* John Wiley & Sons, Inc., 1976
- *Metallography and Microstructures,* Vol 9, *ASM Handbook,* American Society for Metals, 1985
- G. Petzow, *Metallographic Etching,* 2nd ed., ASM International, 1999
- *Sample Preparation of Ceramic Material,* Buehler Ltd., 1990
- W.F. Smith, *Principles of Materials Science and Engineering,* 3rd ed., McGraw-Hill, Inc., 1996
- *Structure,* Vol 33, 1998
- C. Suryanarayana and M.G. Norton, *X-Ray Diffraction, A Practical Approach,* Plenum Press, 1998
- G.E. Vander Voort, *Metallography: Principles and Practice,* American Society for Metals, 1984
- L.H. Van Vlack, *Physical Ceramics for Engineers,* Addison-Wesley Publishing Co., Inc., 1964

CHAPTER 2

Ceramic Fabrication

Ceramics

AN ALL-ENCOMPASSING definition of *ceramics,* from the Greek word *keramos,* meaning pottery or potter's clay, is elusive. Dictionary definitions tend to emphasize pottery, earthenware, decorative tile, and porcelain. Historically, this is accurate, but the space age has developed a wide variety of materials that can be considered ceramics. Additionally, a number of long-established hydrated materials, such as concrete and plaster of Paris, and castable materials, such as glass, are ceramics. In general, *ceramics* are nonmetallic, inorganic, fabricated solids that have covalent or ionic atomic bonds. A few ceramics have metallic or polymeric properties. Silicon semiconductors, silicone, graphite, and diamond can be considered ceramics. Many composites contain ceramics as the discontinuous phase, the matrix, or both.

Modern ceramics include the clay- and other mineral-based materials that are used for pottery, masonry, and glass as well as many compounds that do not occur in nature. The traditional ceramics are mostly hard, brittle insulators. Modern ceramics include some very tough and conductive materials and materials that span a wide range of mechanical, thermal, electrical, optical, and magnetic properties. Some examples of ceramics are:

- Alumina (Al_2O_3) furnace tubes, thermocouple sheaths, spark plug insulators, laboratory crucibles, electronic substrates, grinding media, and washerless faucet seals
- Aluminum nitride (AlN) electronic substrates of high thermal conductivity
- Barium titanate ($BaTiO_3$) capacitors and ultrasonic transducers
- Borosilicate (SiO_2-B_2O_3) Pyrex (Corning, Inc.) high-temperature glassware
- Cadmium telluride (CdTe) semiconductors in solar cells

- Calcium aluminate ($CaAl_4O_7$) refractory cement
- Chromia (Cr_2O_3) refractory bricks
- Diamond (C_{cubic}) abrasives
- Dry ice (frozen CO_2) for sandblasting and temporary refrigeration
- Ferrite (e.g., $MnFe_2O_4$) magnetic particles on computer discs and recording tape
- Gallium arsenide (GaAs) semiconductors in cellular telephones and laser sources in compact disc players
- Garnet ($Al_2O_3 \cdot FeO \cdot 3SiO_2$) sandpaper
- Graphite ($C_{hexagonal}$) lubricants, fibers, motor brushes, crucibles, electrodes, aircraft brakes, and pencil leads
- Hydrated calcium silicate ($Ca_2SiO_4 \cdot nH_2O$) Portland cement
- Lanthanum hexaboride (LaB_6) cathode filaments in electron microscopes
- Lead magnesium niobate ($PbMg_{0.33}Nb_{0.67}O_3$) gas stove igniters
- Molybdenum disilicide ($MoSi_2$) electric furnace elements
- Phosphor (e.g., tin-doped $Sr_2P_4O_7$) particles in color television screens and fluorescent light tubes
- Porcelain (e.g., Al_2O_3-SiO_2-MgO) dishes, decorative tile, electrical sockets, and bathroom fixtures
- Quartz (SiO_2) crystals in wristwatches
- Sialon (Si_3N_4-Al_2O_3 solid solution) cutting tools
- Silica-lime-soda (SiO_2-CaO-NaO) window glass
- Silicon carbide (SiC) electric furnace elements, refractory bricks, sandpaper, and seal rings
- Silicon nitride (Si_3N_4) bearings, nozzles, and turbine engine components
- Silicon semiconductors in integrated circuit chips
- Spinel (e.g., $MgAl_2O_4$) infrared windows and refractory bricks
- Spodumene (Li_2O-Al_2O_3-SiO_3) CorningWare (Corning, Inc.) baking dishes
- Tungsten carbide (WC) tips on drill bits and saw blades
- Zinc oxide (ZnO) varistors, ultrasonic transducers, and topical medicines
- Zirconia (ZrO_2) gemstones, oxygen sensors, solid oxide fuel cells, fiber optic ferrules, electric furnace elements, and cutlery

In addition:

- Silicone (polysiloxane) compounds—polymeric chains of alternating silicon and oxygen atoms—in lubricants, Silly Putty (Binney & Smith Inc.), and high-temperature rubber are ceramics as well as plastics.
- Fiberglass building materials, reinforced concrete, cermets, graphite-epoxy bicycle frames, and magnetic strips for refrigerator doors are composites that contain ceramics.

The ceramic microstructure has several important features that are discussed in greater detail throughout this book. A *crystal* is a solid structure that has long-range atomic order. Gemstones are cut from relatively large crystals. Liquids have only short-range order, and gases have no order. Glass has many of the properties of a solid but only short-range order similar to a liquid and is therefore noncrystalline or amorphous. A *grain* is an individual crystal bonded to other crystals or glass within a much larger solid. The key difference between a sapphire or ruby gem and a 99.8% or higher Al_2O_3 ceramic is that the gem is a crystal, while the ceramic is composed of many grains. Grains of the same composition and structure in a solid constitute a *phase.* Minor constituents in the microstructure are *secondary phases.* The regions where grains intersect are *grain boundaries,* which significantly affect the properties and transformations of ceramics. Voids in the grains and grain boundaries are *pores,* which also have a tremendous effect on the properties of ceramics. Pores are sometimes considered a secondary phase of zero composition. A grain boundary is essentially an interface where the crystallographic axes suddenly change direction, creating a high-energy zone where secondary phases can nucleate and grow. A point on a grain boundary where three grains intersect is a *triple point,* which has some significance in grain-size measurement techniques. Glass occurs as a secondary phase at the triple points of porcelain and many other ceramics. A fracture within a grain or between just a few grains is a *microcrack.*

A naming convention is used for ceramics of metal-oxide composition. The last two letters of the cation name—e.g., *um*—is replaced by *a.* Thus, aluminum oxide (Al_2O_3) is called *alumina,* silicon dioxide (SiO_2) is *silica,* zirconium oxide (ZrO_2) is *zirconia,* magnesium oxide (MgO) is *magnesia,* and so on. Many oxide ceramics occur naturally as minerals, and the mineral names are occasionally used synonymously with the ceramic phases. Thus, alumina is sometimes called corundum or sapphire, silica is also quartz, calcia is also lime, zirconium silicate ($ZrSiO_4$) is also zircon, and so on.

Commercial Fabrication of Ceramics (Ref 1)

One of the first steps in understanding ceramic microstructures is an understanding of the engineered processes and natural phenomena that create microstructures. Most of the ceramics shown and described in this text were fabricated by sintering.

Sintering is the bonding of solid particles into a monolithic structure as a result of the particles being compacted into an approximate final shape, slowly heated to just below the melting temperature, and held at that temperature for several hours. The driving force for sintering is the high surface energy of fine particles, which is lowered by volume and surface diffusion, and vapor and liquid transport of atoms toward the regions of contiguity between adjacent particles. Densification is a natural consequence of sintering—as the particles change shape to increase their contact area with neighboring particles, the voids that comprised the former interparticle spacing are driven toward free surfaces. The loss of voids during sintering results in a volume reduction and therefore a density increase of the compact (Ref 2).

The rate and degree of densification as functions of time, temperature, and pressure are important in sintering. These dependent variables are also affected by mineral, glass, and polymeric additives, gaseous atmosphere, pressure applied before or during heating, temperature rates, initial particle size, and other factors. The ceramic microstructure resulting from a sintering process can contain primary grains, grains of secondary phases, grain boundaries, pores, glass, and cracks, all of which tend to affect the properties and performance of the bulk ceramic component. One of the ceramic engineer's primary responsibilities is to understand and control the sintering variables in order to control the microstructure so as to predict and optimize the properties and performance of ceramic products.

Several manufacturing processes and densification mechanisms are used commercially. The manufacturing processes include uniaxial pressing or dry pressing, hot pressing, cold isostatic pressing (CIP) or isopressing, hot isostatic pressing (HIP), slip casting, tape casting, roll compaction, extrusion, glass-ceramic processing, and reaction bonding. The densification mechanisms include solid-state sintering, liquid phase sintering, transient liquid phase sintering, and reaction sintering.

Uniaxial pressing is the compaction of spray-dried powder (spray drying removes water and prevents agglomeration on drying but leaves the binder in a wet-mixed ceramic powder) between dies that compress in only one dimension and is widely used in the ceramic industry to make such components as tiles, chip carriers, seal rings, and valve components (Ref 3). Cold isostatic pressing is similar to uniaxial pressing, except that the pressure is applied hydrostatically, that is, in all directions by a rubber mold. Spark plug insulators, tapered tubes, and dishes are usually fabricated by CIP (Ref 4). Slip casting uses a porous mold to extract water from a colloidal suspension called slip, quite unlike spray-dried powder. Porcelain items, such as teacups, sanitary ware, and laboratory wares, are often slip cast (Ref 5). Tape casting and roll compaction are both used to make green (not yet sintered) sheets and thin, flat ribbons that ultimately become the substrates for electronic packages (Ref 6). Extrusion is used to make open-ended furnace tubes from plasticized powder forced through a die that shapes the desired cross section (Ref 7). Glass-ceramic processing begins with a completely amorphous component that is shaped in the glass state but heat treated to nucleate and grow crystals after forming. Lithium aluminosilicate CorningWare cookware is manufactured by the glass-ceramic process (Ref 8). Reaction bonding uses a green compact that gains mass and changes composition by reacting with a gas or liquid environment during sintering, such as a tubular graphite-SiC compact that reacts with a pool of molten silicon to densify and form additional SiC (Ref 9). Hot pressing and HIP are analogous to uniaxial pressing and CIP, respectively, except that the pressure is maintained throughout the sintering cycle. Expensive machinery and short tooling life make hot pressing and HIP more expensive than most other processes, but the high cost is often justified by virtually 100% densification that is not easily achieved by "pressureless" processes (Ref 10, 11).

Solid-state sintering is the basic densification mechanism mentioned in the first paragraph of this section, absent of any liquid phases, and applies to nearly all sintered ceramics to a greater or lesser degree. Two key characteristics of solid-state sintering are overall shrinkage of the component and residual porosity (Ref 12). Liquid phase sintering uses a liquid in the voids between particles to enhance sintering kinetics and to affect the properties of the ceramic component. The liquid is formed, typically, either by a eutectic reaction between different phases in the compact or by melting of one of the phases. The liquid is present at the sintering temperature, wets the solid particles, and dissolves one or more solid phases to some degree. Transient liquid phase sintering is a special case of liquid phase sintering where the liquid enhances densification but disappears at some point during the process, either by evaporating, crystallizing, or dissolving into or reacting with the solid phases (Ref 13). Reaction sintering is the mechanism described in the previous paragraph that results in reaction-bonded ceramics such as SiC and Si_3N_4, where densification and a formation reaction occur simultaneously (Ref 14).

The Suggested Reading at the end of this chapter lists several useful books on ceramic fabrication. The Richerson book and *Ceramics and Glasses,* Volume 4 in the *Engineered Materials Handbook* series, provide comprehensive overviews of ceramic fabrication processes and sintering mechanisms. Reed's book describes ceramic powder preparation in considerable detail.

Laboratory Fabrication of Ceramics

Ceramic research often involves a novel composition or sintering process. The initial test of the composition or process is the production of cylindrical pellets. The pellets facilitate evaluation of the microstructure, phases, and density. This section outlines a procedure for fabrication of experimental ceramic pellets, as follows:

1. Start with presized powders, if available. Many ceramic powders are available by the kilogram from laboratory supply companies such as Alfa Aesar, Baxter, Cerac, Cole-Parmer, Fisher, VW&R, and others.

Oxide and even some nonoxide powders may be available from ceramic manufacturers, aluminum producers, and mining companies, although not necessarily in small quantities. Powders can be sized, if necessary, with a ball mill followed by a sieve analysis. A typical starting powder has particles that are approximately 1 to 5 μm in diameter and a surface area of approximately 0.2 to 4 m^2/g. Too much surface area results in oversized grains in the sintered product. Not enough surface area results in weak bonds between particles and high porosity.

2. Weigh the ceramic powders in accurate proportions. For example, take 96 g of Al_2O_3, 2 g of SiO_2, and 2 g of MgO to make an experimental 96% alumina. Pour each powder into a plastic laboratory bottle with enough capacity for at least twice the volume of powder. The excess volume facilitates thorough mixing.
3. Prepare a liquid binder, such as 20 g of polyethylene glycol and 50 drops of Darvan-C deflocculent in 200 mL of distilled water, or polyvinyl butyral. Distilled water alone can be used, but tap water should be avoided, because the dissolved minerals can affect the outcome.
4. The powders may be mixed simply by securing the lid and shaking the bottle until the dry powders are well mixed, but wet mixing gives more uniform and reproducible results. Add the binder solution to the powder, a few milliliters at a time, and stir with a laboratory spatula until a viscous slurry is formed that is slightly more fluid than honey. A paste that can hold a shape similar to modeling clay is too dry.
5. Add alumina grinding pellets, if desired, secure the lid on the bottle, and agitate the mix in a ball mill for approximately 1 h. A large bottle can be turned by itself on the ball mill rollers. A small bottle may need to be turned inside a ball mill jar. The grinding pellets aid in the mixing and break up agglomerates. Faster rotation does not improve mixing! The rotation frequency must be slow enough to allow the contents of the bottle to experience free-fall with each rotation.
6. Stop the ball mill and remove the lid from the bottle. Set the bottle inside a drying oven at 50 to 80 °C. The oven must not be so hot that it evaporates the binder along with the water. Drying in the bottle can take several days but can be accelerated by pouring the slurry into a Teflon-coated (E. I. Du Pont de Nemours & Co.) baking sheet with raised sides, if the drying oven is large enough to hold a baking sheet. Allow the slurry to dry 12 h or more. This step is the manual equivalent of a spray dryer.
7. The dried slurry usually comes out in large chunks that have significant green strength. Crush the chunks with a mortar and pestle or ball mill until the particles pass through a No. 45 (355 μm) sieve. The sieve is optional but yields a more uniform microstructure than unscreened particles.

8. Lubricate a die and rams with a dilute solution of isostearic acid in acetone. Assemble the lower ram and die and fill the die to approximately 80% of its volume with the mixed, dried powder. Insert the upper ram and compress the powder in a laboratory press to approximately 125 MPa for 30 to 60 s. Too high a compaction pressure can fracture the green compact when the compact is ejected from the die. Too low a compaction pressure can result in a weak, low-density sintered pellet. The most common dies have a round cross section, and the compression is parallel to the axis, but dies for production of modulus-of-rupture (MOR) bars and other shapes are also available. Dies and laboratory presses are available from Fred Carver Inc. in Wisconsin. Depths of the Earth Company in Arizona offers laboratory-sized isostatic presses.
9. Set the green compacts on a refractory block or tray inside a cold programmable box or tube furnace. A sheet of platinum foil beneath the green compacts may be desirable, if contamination from the refractory is a problem. Heat the furnace slowly to the sintering temperature, typically ~300 K below the melting temperature of the primary constituent. Some melting temperatures are given in Table 2.1. Hold the furnace at the sintering temperature for the dwell or soak time and cool the furnace slowly back to room temperature. At some temperature, say 500 °C or thereabouts, the rate of natural cooling becomes so slow that the furnace controller can no longer control the cooling rate. For the experimental 96% alumina, the furnace program might be:

   a. Heat at 100 K/h (approximately 1.7 K/min) to 1000 °C in air.
   b. Heat at 200 K/h to 1640 °C.
   c. Dwell (soak) for 4 h at 1640 °C.
   d. Cool at 200 K/h to 500 °C.
   e. End of program (furnace power off).

Table 2.1 Molar mass and melting temperature of some common ceramics

		Melting temperature	
Ceramic	Molar mass, kg/mole	K	°C
Al_2O_3	101.96	2327	2054
Cr_2O_3	151.99	2603	2330
MgO	40.30	3105	2832
SiO_2	60.08	1996	1723
ZrO_2	123.22	2950	2677
$BaTiO_3$	233.20	1889	1616
$3Al_2O_3 \cdot 2SiO_2$	426.05	2023	1750
$MgAl_2O_4$	142.27	2408	2135
TiB_2	69.50	3193	2920
B_4C	55.26	2743	2470
SiC	40.10	3245	2972
AlN	40.99	2790(a)	2517(a)
Si_3N_4	140.28	2151(a)	1878(a)

(a) Sublimates or melts incongruently at the indicated temperature. Source: Ref 15

10. When the furnace has cooled to room temperature, the pellets or bars are ready for ceramographic analysis. Modulus-of-rupture bars require precision machining prior to mechanical testing. When performing an experiment, keep a record of each step of the fabrication process, including the mixing conditions, binder composition, compaction pressure, heating and cooling rates, soak time and temperature, furnace atmosphere, and any other pertinent details.

REFERENCES

1. R.E. Chinn, "Transient Glass-Phase Processing of $SrAl_2Si_2O_8$-Alumina Ceramics," Thesis T-4685, Colorado School of Mines, 1995, p 43–46
2. R.M. German, Fundamentals of Sintering, *Ceramics and Glasses,* Vol 4, *Engineered Materials Handbook,* ASM International, 1991, p 570–579
3. B.J. McEntyre, Dry Pressing, *Ceramics and Glasses,* Vol 4, *Engineered Materials Handbook,* ASM International, 1991, p 141–146
4. F. Kennard, Cold Isostatic Pressing, *Ceramics and Glasses,* Vol 4, *Engineered Materials Handbook,* ASM International, 1991, p 147–152
5. C.H. Schilling and I.A. Aksay, Slip Casting, *Ceramics and Glasses,* Vol 4, *Engineered Materials Handbook,* ASM International, 1991, p 153–160
6. R.E. Mistler, Tape Casting, *Ceramics and Glasses,* Vol 4, *Engineered Materials Handbook,* ASM International, 1991, p 161–165
7. I. Ruppel, Extrusion, *Ceramics and Glasses,* Vol 4, *Engineered Materials Handbook,* ASM International, 1991, p 166–172
8. L.R. Pinckney, Phase-Separated Glasses and Glass-Ceramics, *Ceramics and Glasses,* Vol 4, *Engineered Materials Handbook,* ASM International, 1991, p 433–438
9. Reaction-Forming Processes, *Ceramics and Glasses,* Vol 4, *Engineered Materials Handbook,* ASM International, 1991, p 236–241
10. A. Ezis and J.A. Rubin, Hot Pressing, *Ceramics and Glasses,* Vol 4, *Engineered Materials Handbook,* ASM International, 1991, p 186–193
11. H.T. Larker, Hot Isostatic Pressing, *Ceramics and Glasses,* Vol 4, *Engineered Materials Handbook,* ASM International, 1991, p 194–201
12. M.F. Yan, Solid-State Sintering, *Ceramics and Glasses,* Vol 4, *Engineered Materials Handbook,* ASM International, 1991, p 270–284
13. O.-H. Kwon, Liquid-Phase Sintering, *Ceramics and Glasses,* Vol 4, *Engineered Materials Handbook,* ASM International, 1991, p 285–290
14. J.S. Haggerty, Reaction Sintering, *Ceramics and Glasses,* Vol 4, *Engineered Materials Handbook,* ASM International, 1991, p 291–295
15. M.E. Schlesinger, Melting Points, Crystallographic Transformation, and Thermodynamic Values, *Ceramics and Glasses,* Vol 4, *Engineered Materials Handbook,* ASM International, 1991, p 883–891

SUGGESTED READING

- *Ceramics and Glasses,* Vol 4, *Engineered Materials Handbook,* ASM International, 1991
- J.S. Reed, *Principles of Ceramic Processing,* 2nd ed., John Wiley & Sons, Inc., 1995
- D.W. Richerson, *Modern Ceramic Engineering: Properties, Processing, and Use in Design,* 2nd ed., Marcel Dekker, Inc., 1992

CHAPTER 3

Sawing and Mounting

THE PURPOSE of sawing is to extract a representative sample from a ceramic component that is suitable for polishing and microscopic examination or to isolate a region of interest. The purpose of mounting is to encapsulate, impregnate, or embed the section in a plastic disc of appropriate size and composition that facilitates preparation and identification. These are usually the first steps in ceramography. The sawing and mounting operations must always be done so as to minimize the damage to the specimen.

Sawing

Saw the ceramic component with a water-cooled, low-density, metal-bonded diamond wafering blade on a high-speed cut-off machine, as shown in Fig. 3.1(a). A resin-bonded diamond blade may be advantageous for some specimens. Use a load of approximately 5 to 10 N (500 to 1000 g_f) and a blade rotation rate of 2000 to 5000 rpm for dense ceramics. Use 1 N and 500 rpm for most refractories, concrete, and semiconductors. A low-speed saw (Fig. 3.1b) that uses oil or kerosene as the coolant and lubricant can also be used, although the cutting rate can be excruciatingly slow for dense ceramics at 500 rpm. Both types of saw are available from several of the manufacturers listed in Appendix B. Feed the component into the blade (or the blade into the component) slowly but steadily to avoid jagged sawed surfaces. Remove the burrs, if necessary, with a coarse (400-grit or 45 μm) diamond grinding wheel.

The best sawed surfaces come from an automatic saw equipped with a double saddle chuck. The double saddle chuck (Fig. 3.2a) prevents burrs by holding both halves of the component firmly until the cut is complete. A specimen held in a single-clamp chuck usually fractures when the blade is approximately 90% through the cross section, and that fracture is almost never in the plane of the kerf. If a double saddle chuck cannot be used, the free half (on the opposite side of the blade from the clamp) of the specimen is easy to find in the bottom of the coolant reservoir if an electrician's alligator clip is attached to the free half. Secure the alligator clip to the saw cabinet via a braided wire, as shown in Fig. 3.2(b).

Sawing should always be done in a way that minimizes damage to the specimen, such as overheating, dislocations, deformation twinning, and cracking on or near the sawed surface. While ceramics are usually not as susceptible to overheating damage as most other materials, a liquid coolant should always be used. The ceramic must be insoluble in the coolant. Some ceramics and minerals have water-soluble phases, for example, CrO_3 in

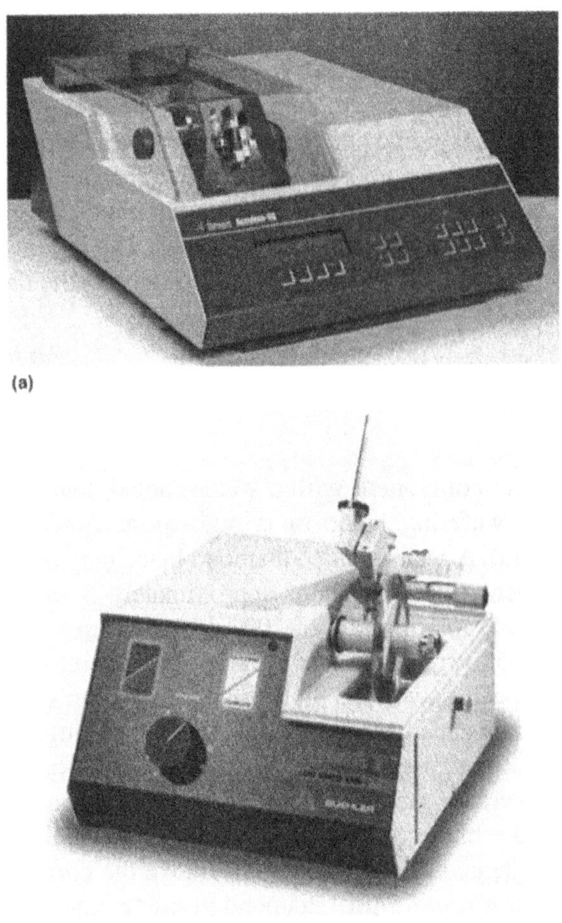

(a)

(b)

Fig. 3.1 Cut-off machines for sawing ceramographic specimens. (a) High-speed machine. The specimen is fed into the diamond wafering blade at a constant rate and force that are controlled electronically. (b) Low-speed machine. The specimen is fed into the diamond wafering blade by the force exerted by a deadweight. The blade rotation rate is fixed at 500 rpm.

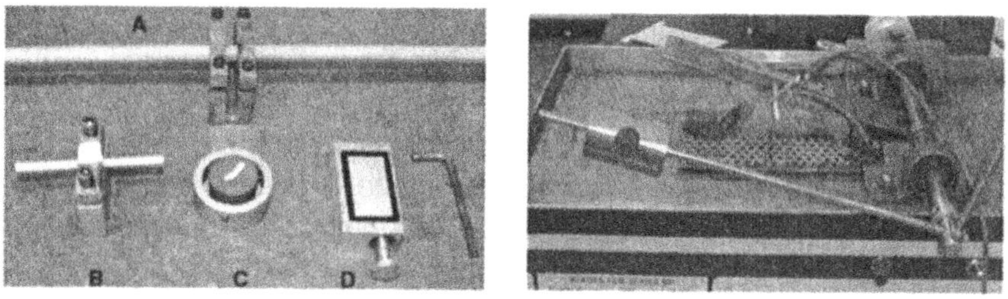

(a) (b)

Fig. 3.2 Chucks for a cut-off machine. (a) A, double saddle chuck holding a ceramic tube; B, standard (single saddle) chuck; C, chuck for round objects, such as the mounted specimen; D, thin-section vacuum chuck. (b) A tethered alligator clip prevents the unsecured piece from being lost in the coolant trough when used with a single-clamp chuck. Courtesy of A.H. Hunt

some Cr_2O_3 refractories, and should be sawed in kerosene or alcohol. Blades other than metal-bonded diamond can be used, but the amount of damage caused by the blade should always be considered. Damaged blades should not be used. A clamped component is less likely to be damaged during sawing than a hand-held component and is less likely to cause injury to personnel. To increase the cutting rate, increase the blade rotation rate rather than the force. A higher force is more likely to damage the specimen than a higher blade rotation rate. The effects of cutting speed, force, and blade type on the as-sawed surface of several ceramics have been compared (Ref 1).

Saw the surface to be examined to a diameter that is 10 to 50% of the diameter of the mold to be used. A 5 to 10 mm diameter specimen, 5 to 10 mm high, in a 25 or 30 mm mold is usually a good fit. Friable ceramics should be mounted in castable resin first, as described in the section "Mounting" in this chapter, then sawed.

In some cases, a fracture surface is necessary or desirable. To make a fracture surface, notch the component with the diamond wafering blade where the fracture should be initiated. Clamp the component in a vise with the notches at or just above the top of the vise jaws and strike the component with a hammer. This technique is useful when the component is too large to fit into the saw or when an etching method is unavailable.

Other methods have been used to cut ceramics successfully, including laser beams, supersonic water jets, and reciprocating saws. The machinery for these techniques is not widely available in a configuration that is suitable for a ceramographic lab, which is why they are not described here.

When sawing is complete, clean and dry the specimen, as described in Subroutine 3.1.

Subroutine 3.1: Cleaning and Drying Sawed Specimens

Cleaning. If an oily sawing lubricant is used, remove the lubricant by putting the specimen in a beaker of warm water with detergent, such as Ivory (Proctor & Gamble Co.), Micro-90 (International Products Corp.), or Alconox (Alconox, Inc.). Use distilled or deionized water if the tap water is too hard. Be sure the specimen is not water soluble. Put the beaker in the basket of an ultrasonic bath and fill the tank with enough warm water so that the beaker is wetted on the outside but not buoyed. Turn on the ultrasonic bath for 30 to 120 s or until loose debris is removed from the specimen. Remove the specimen from the beaker with a tweezer and rinse it in warm water. Most ultrasonic bath manufacturers recommend the use of a basket to suspend the beaker in the tank rather than setting the beaker on the bottom of the tank. This process of ultrasonic cleaning is *ultrasonication.*

Drying. Squirt the wet specimen with a volatile solvent, that is, a drying agent, such as reagent-grade ethanol, acetone, or isopropanol. Hold the specimen in a stream of warm air from a heat gun or specimen dryer until all the liquid evaporates. Clean, compressed air can be used in lieu of the heat gun but beware of plant air systems that contain oil mist to lubricate pneumatic equipment. Aerosol dust removers, such as Dust-Off (Falcon Safety Products) or Beseler DustGun 100, can also be used. A hot plate or drying oven at 80 to 120 °C can also be used to dry specimens but could have an adverse effect on hydrated ceramics (e.g., concrete) and some minerals.

Mounting

Thick sections can be mounted in either of two ways. Strong, fully densified ceramics are best mounted in powdered resins that polymerize while heated in uniaxial compression. The powdered resins may contain filler particles that enhance certain properties of the cured resin. Older mounting presses resemble car jacks (Fig. 3.3a), while newer mounting presses (Fig. 3.3b) use water cooling and compressed air to greatly increase the production rate. Friable, fragile, or highly porous ceramics are best mounted in castable, liquid resins that polymerize with the addition of a catalyst or hardener.

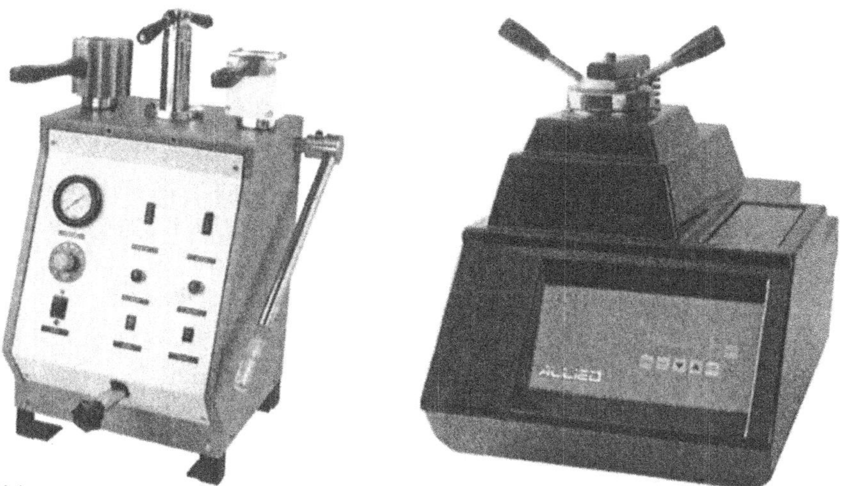

(a) (b)

Fig. 3.3 Mounting presses. (a) Old-style mounting press. The dies are cooled by a manually applied metal heat sink. (b) Automatic mounting press. The pressure and heating and cooling times are controlled automatically. The dies are water cooled.

Mounting resins are often listed in metallographic catalogs as consumables. Consumable refers to accessories that are used once or only a few times, such as abrasives, polishing cloths, and mounting resins.

Compression Mounting Method—Automatic Mounting Press. The steps are:

1. Lubricate the dies with mold-release aerosol or wax. Use the wax sparingly, a few milliliters at a time.
2. Be sure the specimen is clean and dry. Place the flat surface of interest downward and in the center of the lower ram. Use a plastic or metal clip (available from most equipment and consumables vendors) to support unsteady specimens such as thin plates on edge. Lower the ram 35 to 40 mm into the outer die.
3. Fill the mold to the top of the specimen with mineral-filled epoxy resin or glass-filled diallyl phthalate resin, as shown schematically in Fig. 3.4 (Ref 2). The resins are described in Table 3.1.
4. Fill the outer die to 5 mm below the rim with phenolic resin or a phenolic premold disc. Level the phenolic powder by tamping it with the upper ram. Phenolic is available in several colors and can be used to color code ceramographic specimens. Phenolic contains wood flour filler particles and is less expensive and less abrasion resistant than the two resins specified in step 3.
5. Center a paper label, if desired, atop the leveled phenolic. Business card stock, cut into ~15 by ~20 mm rectangles, makes better labels

than writing paper does and is a good way to recycle obsolete business cards.

6. Fill the outer die to 2 mm below the rim with unfilled acrylic powdered resin, which is transparent on polymerization. Be sure the paper label is covered completely and uniformly.
7. Place the upper ram gently on top of the acrylic resin, slowly lower the ram until the threaded part of the upper ram is seated on the threads of the outer die, and thread the upper ram onto the outer die. Loosen the upper ram by 1/4 turn from the "tight" position. Raise the lower ram to compress the mold assembly to the pressure specified by the press manufacturer, typically 28 MPa for a 30 mm die. Note that the pressure shown on the gage is usually the air pressure supplied to the press, not the die compression. Table 3.2 shows the amount of force needed for each die diameter. Apply heat to the mold assembly and hold the temperature at 150 °C for 5 to 10 min or as recommended by the resin manufacturer.
8. Cool the mold assembly under pressure to below 40 °C when the resins are fully cured.
9. Lower the pressure so that the upper ram can be removed, and then raise the ram again to eject the encapsulated specimen.
10. Thoroughly clean the flash out of the dies before making another mount.

Compression Mounting—Hydraulic Mounting Press. The procedure for compression mounting in a conventional hydraulic mounting press is the same as for the automatic mounting press mentioned previously,

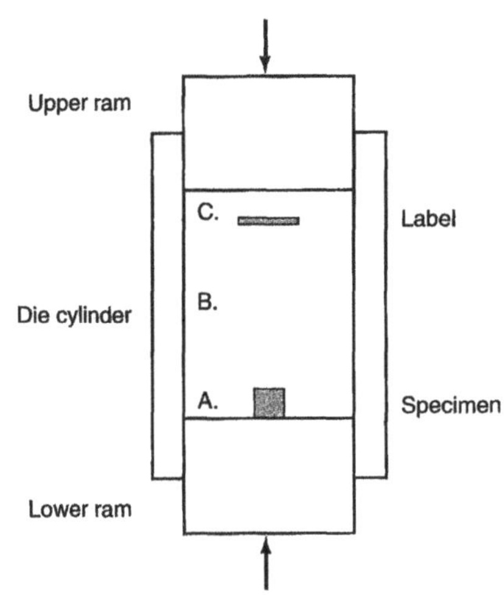

Fig. 3.4 Cross section of a loaded compression die. Region A is mineral-filled resin surrounding the specimen, B is flour-filled phenolic, and C is unfilled acrylic resin above the paper label (Ref 2).

Table 3.1 Compression mounting resins

Plastic	Filler	Trade names	Characteristics	Application
Phenolic	Wood flour	Bakelite, MultiFast, Phenocure	Thermosetting, opaque, inexpensive, low abrasion resistance, available in several colors	Encapsulation filler
	Metal or carbon particles	Konductomet, PolyFast, Probemet	Thermosetting, opaque, conducts electricity	Encapsulation filler for electrolytic polishing and etching or examination by scanning electron microscopy (SEM)
Diallyl phthalate	Glass fiber	IsoFast	Thermosetting, opaque, medium abrasion resistance	Polished face of hard materials
	Mineral	...	Thermosetting, opaque, high abrasion resistance	Polished face of ceramics and minerals
Epoxy	Mineral	Epomet, DuroFast	Thermosetting, opaque, high abrasion resistance	Polished face of ceramics and minerals
Acrylic	None	Transoptic, Lucite, SpeciFast, Plexiglas	Thermoplastic, transparent, low abrasion resistance	Encapsulation filler for embedded labels
	Iron	ConduFast	Thermoplastic, conducts electricity	Encapsulation filler for electrolytic polishing and etching or examination by SEM

except that the mold assembly is compressed between the movable lower platen and the fixed upper platen of a hydraulic press and the die assembly is cooled by a heat-sink jacket rather than water.

Be sure the specimen is clean and dry. Place the specimen on the lower die—the shorter of the two ram dies—flat side downward and in the center of the die. Gently seat the mold cylinder around the lower die. Fill the mold cylinder as described previously in steps 3 to 6 and in Fig. 3.4. Place the upper die—the longer of the two ram dies—gently in the top of the outer die cylinder and press it down until in touches the acrylic resin. Set the entire mold assembly in the center of the lower platen of the hydraulic press. Place the heating device around the mold assembly and switch on the heat. Pump the hydraulic press up to 28 MPa and insert the dial thermometer through the upper platen into the mold assembly. Table 3.2 shows the amount of force needed for each die diameter. Hold the mold temperature at 150 °C for 5 to 10 min or as recommended by the resin manufacturer. Remove the heater and apply the cooling device when the resins are fully cured. Allow the mold to cool to below 40 °C. The cooling rate can be moderately accelerated by storing the cooling device in a freezer until it is needed. Eject the mount by pressing the upper ram all the way through the outer die cylinder.

Shrinkage on cure of the mount is approximately 50% of the height of the uncompacted resin. Thus, a mold that is 40 mm deep prior to compaction yields a mount that is approximately 20 mm high. The mineral-filled epoxy most closely matches most ceramic specimens in terms of abrasion resistance and therefore results in the best edge retention during grinding. A metal- or carbon-filled resin could be used in place of the wood-filled phenolic, if the specimen is to be examined by scanning electron microscopy or electrolytically etched. The transparent acrylic resin

Table 3.2 Volumes and areas of molds for castable mounts

Mold diameter		Mold volume		Face area		Force(a)	
mm	in.	mL	oz	mm²	in.²	kN	lb
25	1.0	13	0.5	491	0.79	13–14	3100
30	1.25	20	0.7	707	1.23	19–20	4900
40	1.5	35	1.2	1260	1.77	35	7100

(a) The amount of force to apply to a mounting press to achieve a compressive stress of approximately 28 MPa (4000 psi)

can be omitted and that space filled instead with more phenolic. The mount can be identified with an engraving pen or adhesive label.

Mounts that crack or otherwise fail during polymerization have several possible causes, as shown in Fig. 3.5 (Ref 3). Cracks could be caused by thermal expansion mismatch between ceramic and resin; moisture in the resin; insufficient time, temperature, or pressure for resin cure; excess cure temperature; an oversized specimen; or sharp corners on the specimen. An unfused mount or acrylic resin that is not transparent could be caused by insufficient pressure, temperature, or cure time. Premature ejection of a hot mount can cause edge shrinkage or cracking. Overly hot dies at the start can cause failure through nonuniform polymerization of the resins.

Castable Mounting. Although powder resins cured in a hot press yield the best ceramographic results in most applications, the castable liquid resins have several applications. The liquids penetrate and fill the pores and other voids of ceramics that are not fully densified, such as refractory bricks, especially with the aid of a vacuum chamber for mold filling and a pressure vessel for curing. The low-stress casting process preserves friable ceramics, mineral and slag sections, plasma- or flame-sprayed coatings, and fragile microelectronic devices. Fragile and friable ceramics can be mounted in castable resin prior to sawing, in order to minimize destruction by the saw blade.

Castable mounting is also known as cold mounting. This is somewhat of a misnomer, due to the exothermic polymerization reaction. The maximum temperature in the liquid resins is typically approximately 120 °C, less than the 150 °C of the compression method. The unfilled liquid resins have relatively low abrasion resistance but can be enhanced by the addition of ceramic particles.

Some resin-curing agents are quite hazardous and should be used with caution. A fume hood may be required to exhaust poisonous vapors. Protective clothing is recommended, especially gloves, to prevent skin irritation. Most resins are flammable and must be kept away from sparks, flames, and other heat sources. The National Fire Protection Association diamond symbol on the container label gives a brief description of the hazards, while the MSDS provides a more thorough description of hazards and treatments.

Defect		Cause	Solution
Radial split	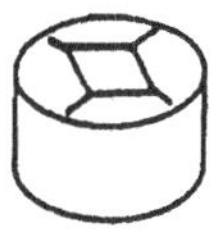	1. Specimen too large for mount 2. Specimen has lower thermal expansion than mounting medium 3. Excessively sharp specimen corners	1. Use larger mold or smaller specimen 2. Use casting method or other thermosetting resin 3. Use casting method
Edge shrinkage	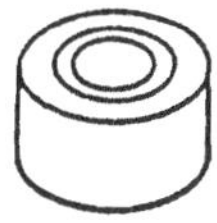	1. Specimen has higher thermal expansion than mounting medium	1. Use casting method or other thermosetting resin
Burst (bottom) Burst (top)	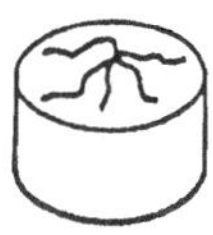 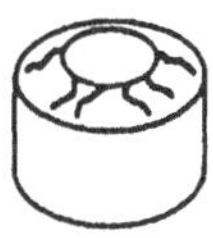	1. Moisture in or on specimen 2. Moisture in resin 3. Insufficient curing time and/or pressure	1. Dry specimens thoroughly 2. Dry resin in preheater 3. Increase curing time and/or pressure
Unfused	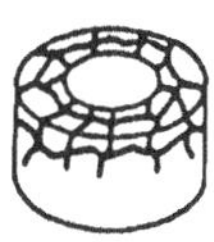	1. Premature curing of resin 2. Insufficient molding pressure 3. Insufficient mold temperature	1. Cool die before adding resin 2. Apply pressure immediately after adding resin 3. Increase mold temperature
Circumferential split	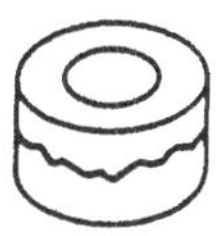	1. Moisture in resin 2. Mold initially too hot 3. Mold temperature too high	1. Dry resin in preheater 2. Cool die between mounts 3. Reduce mold temperature

Fig. 3.5 Defects in compression resins (Ref 3)

Castable mounting method steps are:

1. Lubricate the casting mold cup with liquid mold release. Do not use the mold release on the mold base yet—the mold release prevents the adhesive tape from adhering. Several molds can be prepared at one time, depending on the volume of resin needed to fill all the molds.
2. Adhere double-stick (adhesive on both sides) tape to the center of the upper surface of the casting mold base, as shown schematically in Fig. 3.6 (Ref 2). The adhesive tape is optional but prevents the specimen from moving or floating when the resin is poured into the mold. Daub the mold release on the base around the tape.
3. Be sure the specimen is clean and dry. Peel off the tape backing and adhere the specimen to the exposed side of the tape. Securely place the mold cup over the mold base. The surface to be polished is downward, in contact with the tape.

4. Mix the resin and catalyst in a paper cup according to the manufacturer's instructions—typically, two parts powder to one part liquid by mass or volume for acrylic. Use a wooden or steel spatula to stir but avoid agitating the resin mixture to the point of aeration. Gently fill the mold to 3 to 5 mm below the rim, as in Fig. 3.6. The volumes of common mold sizes are given in Table 3.2. For improved results, fill the mold in a vacuum chamber at approximately −80 kPa gage (10 to 25 kPa, or 80 to 200 torr absolute) pressure so that air bubbles are less likely to be trapped in the pores by the resin. Several equipment manufacturers offer a vacuum chamber that is specifically designed for this purpose.
5. If a friable or fragile specimen is embedded to protect it during sawing, use only enough mixed resin to cover the specimen. After sawing, trim the plastic around the specimen as needed and remount the specimen in either compression or castable resin.
6. Gradually return the vacuum chamber, if used, to atmospheric pressure. Immediately place the filled mold into a pressure vessel, seal the vessel, and increase the pressure to 2 to 3 atm. Allow ample time for the resin to cure completely. Acrylic resin cures in as little as 10 min, but some epoxy resins require several hours and can be left in the pressure vessel overnight. The pressure vessel is also optional, but it removes bubbles and decreases the opacity of the cured resin. The polymerization rate of most epoxy and polyester resins can be doubled or tripled in a drying oven at 70 to 80 °C, compared to the room-temperature rate. The cure characteristics of castable resins are listed in Table 3.3.
7. If a more abrasion-resistant face is desired for improved edge retention, pour a few milliliters of the fresh-mixed resin into another cup and add enough alumina granules to form a slurry. Stir the granules into the resin and pour the slurry into the mold first, to a depth of no more than 5 mm. Pour the remaining unfilled resin over the slurry to near the rim of the mold and continue as in step 4. Some castable resins are now available with mineral filler.
8. Dyes, such as Struers AcryDye are available for color coding the encapsulation. Fluorescent dyes, such as EpoDye and Fluor-Met, (Buehler, Ltd.) are available that make pores and cracks more visible in a microscope equipped with a specialized filter.
9. An alternative to the reusable cup mold is the one-time-only ring mold. Several metallographic manufacturers offer phenolic rings. A section of plastic or metal pipe can also be used. Plastic pipe can be knurled on the inner surface and primed with methyl ethyl ketone for improved adhesion to the resin. Set the ring mold on a smooth surface, such as a pane of glass. You may need to apply foil and masking tape around the bottom edge of the ring mold to prevent the liquid resin from leaking. Apply mold release to the glass. Center the specimen in

the ring and fill the mold as in steps 2 to 6. The advantage of the ring mold over the reusable cup is that the ring mold does not shrink in the radial direction on curing.

10. If the mold leaks, fill it to only 5 mm depth or less. When the initial resin cures and seals the leak, fill the remainder of the mold with a fresh batch of the same type of resin.

Use powder resins for most ceramographic applications. Mineral-filled powders have higher hardness and abrasion resistance than most other powder and liquid resins and result in better edge retention. Use castable liquid for the following applications:

Material	Example
Highly porous ceramics	Refractory brick, some porcelains, concrete, minerals
Fragile components	Microelectronic devices, cerdips
Hollow components	Tubes, channels, thermocouple sheaths
Thin plates	Substrates for microelectronic devices on edge. Support the plate on edge with a plastic or steel clip, available from most metallographic vendors.
Fast mounting	Up to approximately ten 25 mm (1 in.) molds simultaneously
Coated surfaces	Metallized, flame-sprayed, plasma-sprayed, glazed and plated devices; porcelain enamel

Some other procedures for ceramographic preparation are listed in Appendix A. Among these are ASTM B 665, C 994, and E 3. Castable resins can also be used for replication, as discussed in the section "Replication and Field Ceramography in Chapter 7.

Table 3.3 Cure characteristics of castable resins

Polymer	Selected trade names	Cure time at 20–25 °C	Cure time at 70–80 °C
Acrylic	Sampl-kwick, Fast Cure, Koldmount, Quickmount, Klearmount, Lecoset 7007, AcryFix, CitoFix, DuroFix, Varidure	5–55 min	Not specified
Epoxy	Epo-kwick, Extec Fast Cure, Mager Fast-Set, Epofix, SpeciFix, CaldoFix, Epoxide, Epo-mix, Epoxicure, Epo-Color	30 min to 20 h, depending on brand	15 min to 2 h, depending on brand
Polyester	Castolite, Castoglas, Lecoset 7000, SeriFix, TrioFix	6 min to 4 h, depending on brand	20–30 min

Note: DuroFix and TrioFix both contain mineral fillers. Most others are unfilled.

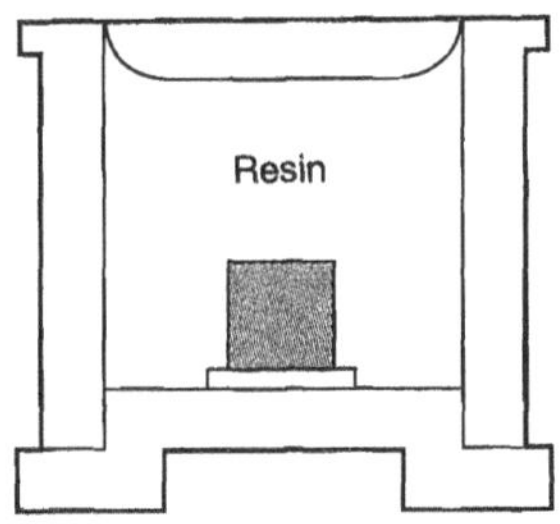

Fig. 3.6 Cross section of a filled mold for castable resin (Ref 2). The specimen is held in place by double-stick tape. The free surface (top) of the resin is not flat, due to the meniscus where the resin is attracted toward the mold cup by capillary action.

Edge Retention

The edges of polished faces tend to be rounded, especially when the encapsulation has a significantly different resistance to abrasion. This lack of edge retention can be ruinous to small surfaces, including plates mounted on edge. One cannot focus on rounded edges at higher magnifications with a microscope. The interface between phases is obscured when one of the edges is rounded.

One way to retain edges is to match the encapsulation with the specimen, such as mineral-filled epoxy with dense ceramics, as already mentioned. The encapsulation can be modified in some cases, such as ceramic granules or steel shot embedded in liquid resin. Another way is to protect the edges of interest with edges that can be sacrificed. The sacrificial edges can and should be used in addition to the matched or modified resin.

Metallographers sometimes use nickel plating to support edges. The plating gets rounded during grinding and polishing instead of the important edge. In ceramography, a thin plate can be sandwiched between two similar plates. The three plates can be held together with adhesive tape, squeezed by a plastic or steel clip (available from most vendors of mounting resins), or glued with the thin-section epoxy mentioned in Chapter 6.

Another problem that can occur at the specimen-encapsulation interface is shrinkage by the encapsulation. On curing and cooling, the encapsulation shrinks away from the specimen, due to differences in thermal expansion, and leaves a gap at the interface. The gap acts as a trap for swarf and abrasive particles and interferes with subsequent grinding and polishing steps. The shrinkage is minimized if the compression resin is cooled to room temperature while under pressure, which is fast and easy with an automatic mounting press. The effect may be more difficult to control with castable resins.

Beveled Edge

A beveled or rounded edge on the encapsulation surface to be polished, as in Fig. 3.7, may extend the life of polishing paper. An initial edge radius of 2 to 3 mm is ideal. A metal-turning lathe can be used to create a very uniform bevel. Some mounting press manufacturers offer a mold insert that creates the bevel or radius at the start, thus eliminating the need for postcure machining.

To bevel an edge without a lathe, hold the mount on a coarse grinding disc or belt with the axis of the mount approximately 45° to the plane of the abrasive. Rotate the mount slowly about its axis to grind away the edge. Continue to rotate the mount until the bevel is 2 to 3 mm high all the way around the mount.

Ceramographic Mounting Resins

Plastic resins are used to encapsulate ceramic sections to facilitate identification and materialographic preparation. The materials science aspects of plastic processing are described briefly in this section to aid the materialographer in the correct choice and use of these plastics. All the plastics described here are organic polymers, that is, thousands of carbon atoms linked together in a chain or network, with other atoms attached sideways along the molecular chain. Silicones, primarily used as lubricants and high-temperature rubber, are inorganic polymers.

Resin refers to plastic raw materials, as uncured monomer liquids or partially cured oligomers in pellet form. Monomers are the basic molecular units of plastics, such as an ethylene or ester molecule, which are usually sold as liquid solutions. A few (≤25) monomers linked together make an oligomer or prepolymer. One thousand or more monomers linked together to form a long chain or network become polymers, such as polyethylene or polyester. The exothermic linking or curing reaction is polymerization, which can be as fast as a few minutes or as slow as several hours, depending on temperature and other conditions. *Plastics* are solids that contain one or more polymers and sometimes a filler material

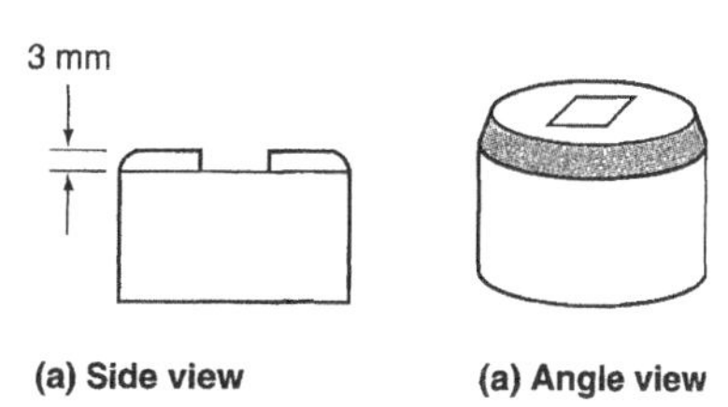

Fig. 3.7 Beveled edge. A beveled edge on the encapsulation is less likely to snag and tear a polishing paper than a square edge.

that gives the plastic certain properties. Filled plastics are also called composites or reinforced plastics.

The curing process may be initiated by a catalyst, heat and pressure, or a chemical reaction with a hardener. A cured plastic generally has both amorphous and crystalline regions, with the degree of crystallinity dependent on the monomers and the processing conditions. Vulcanized rubber and polyvinyl chloride (PVC) are almost entirely amorphous, while high-density polyethylene is approximately 95% crystalline. Polymers that can be melted or dissolved and reformed, due to a linear molecular structure, are *thermoplastics.* Acrylic, nylon (polyamide), PVC, polystyrene, and polyethylene are thermoplastic. Highly cross-linked polymers degrade irreversibly when overheated or chemically attacked and are called *thermosets.* Epoxy, phenolic, diallyl phthalate, polyurethane, and vulcanized rubber are all thermosetting. Most of the ceramographic resins, except acrylic, are thermosetting.

Solid thermosetting resins are fully cured in a three-stage process. In the A-stage, two organic compounds, such as an alcohol and an acid, react to form an oligomer or prepolymer that remains soluble and fusible, to some degree. Mineral, glass, metal, or wood filler particles are added while the oligomer is in solution. In the intermediate B-stage, the solvent is removed to form a solid called resitol that can still be molded and fused with the application of heat and pressure. Thermosetting compression mounting resins are sold as pellets in the B-stage. The final C-stage is the fully cured, fused, cross-linked resite that has been molded under heat and pressure. Fully cured thermosets can be removed from the mounting press while still warm, due to low thermal expansion.

Engineering Plastics, Volume 2 in the *Engineered Materials Handbook* series, provides a comprehensive overview of plastic materials and processes.

Acrylic is a family of thermoplastics that includes polymethyl methacrylate. An initiator such as benzoyl peroxide or 2-2'-AIBN in solution, forms a free radical (unpaired electron) that attacks the double bond in a methyl methacrylate monomer. Each monomer becomes a new free radical and lowers its energy by bonding with similar monomers, a process called free-radical vinyl polymerization (Ref 4). Acrylics should be allowed to cool to approximately room temperature before removal from the mold, whether compression mounted or cast. Cast acrylics generate considerable heat during polymerization.

Acrylic resins are used for windows, paint, bathtubs, and lubricants. The high transparency and toughness of acrylic make it a suitable replacement for glass in high-stress optical applications. In materialography, unfilled acrylics are used where a transparent encapsulation is needed, but high abrasion resistance is not required. Castable acrylic, because of its fast curing time at room temperature, is used where a high throughput rate is required.

Diallyl phthalate (DAP) orthoresins are thermosetting polyesters that result from a reaction between allyl alcohol and phthalate anhydride. The monomers are partially polymerized in the liquid state with the addition of a free-radical initiator and fillers and heated to form a fusible prepolymer solid (Ref 5). Diallyl phthalate is used to make electrical insulators, such as serial plugs for computer peripherals. In materialography, mineral- or glass-filled DAP is used where high abrasion resistance and low shrinkage are needed, such as surrounding the surface to be polished of ceramics and other hard materials.

Epoxy and Epoxide. Epoxy resins begin with epoxide ring prepolymers that are derived from epichlorohydrin. The liquid epoxide is cured with the addition of a liquid hardener. Active groups on the hardener—a diamine monomer—open the reactive epoxide rings to produce cross links (Ref 6). The epoxy polymerization reaction produces no byproducts, resulting in low shrinkage. Epoxies are used for adhesives that can bond to metal and as the matrix in fiber-reinforced composites.

In materialography, mineral-filled epoxy is used where high abrasion resistance, good chemical resistance, and low shrinkage are needed, such as surrounding the surface to be polished of ceramic sections. Mineral-filled epoxy is the resin of choice for most ceramographic applications. Unfilled, castable epoxy usually has low shrinkage and is used for friable or fragile specimens or materials that have a continuous network of pores.

Phenolic is also called Bakelite after its inventor, Leo Baekeland, who created the world's first totally synthetic polymer circa 1905. Phenolic, a thermoset, is the result of a condensation reaction between phenol (C_6H_5OH) and formaldehyde (HCHO) (Ref 6). A water molecule is evolved by each monomer in a condensation reaction.

In the A-stage of phenolic production, phenol and formaldehyde react to form a thermoplastic, low-molecular-mass prepolymer called resole. Wood flour or electrically conductive particles, colorants, and a catalyst are added to resole dissolved in a solvent. The filled resole is dried to form the partially cured pellets (B-stage) that are sold by materialographic vendors. The application of heat and pressure in the metallographic mounting press softens the pellets enough to cause flow and bonding and completes the polymerization (C-stage). Water and formaldehyde vapors may evolve (outgas) from the reaction. The result is a hard, stiff, opaque cross-linked plastic that can be ejected from the mold while still hot. The prepolymer may also be called novolac, depending on the ratio of phenol to formaldehyde.

Phenolic resins are used as plywood adhesives, bottle caps, radio knobs, electrical insulation, and cookware handles. Wood-filled phenolic has very low thermal and electrical conductivity and is used as an inexpensive, color-coded filler away from the polished surface of harder materials. Metal- or carbon-filled phenolic is electrically conductive and is used in materialographic applications that include electrochemical etching and scanning electron microscopy. Phenolic preform discs are used to increase mounting throughput rate.

Polyesters are the result of condensation reactions between alcohols and organic acids, such as glycol and dicarboxylic acid. Polyesters can be thermoplastic, such as polyethylene terephthalate and polycarbonate, or thermosetting, such as DAP. The uncured thermoset resins are unsaturated, that is, have some double bonds between carbon atoms. The addition of a monomer, such as styrene, and a catalyst opens the double bonds in the polymerization reaction and forms a cross-linked, rigid network (Ref 6). Polyesters are used for textiles, beverage containers, sleeping bag insulation, lenses, optical filters, and recording tape. In materialography, unfilled polyesters may be used interchangeably with castable epoxy and acrylic resins, where high abrasion resistance is not needed.

REFERENCES

1. D.C. Zipperian, S. Chanat, and A. Trujillo, Ceramic Microstructural Analysis for Quality Control, *Am. Ceram. Soc. Bull.,* Vol 71 (No. 7), July 1992, p 1077–1098
2. R.E. Chinn, Preparation of Microstructures of Alumina Ceramics, *Structure,* Vol 33, 1998, p 16–20
3. H. Mörtel, Microstructural Analysis, *Ceramics and Glass,* Vol. 4, *Engineered Materials Handbook,* ASM International, 1991, p 570–579
4. M. Michalovic, G. Brust, and D. Parrish, *The Macrogalleria,* Department of Polymer Science, University of Southern Mississippi, 1995–2000
5. M.R. Greer, Allyls (DAP, DAIP), *Engineering Plastics,* Vol 2, *Engineered Materials Handbook,* ASM International, 1988, p 226–229
6. N.G. McCrum, C.P. Buckley, and C.B. Bucknall, *Principles of Polymer Engineering,* Oxford University Press, 1988, p 258, 259, 373–374

SUGGESTED READING

- *Engineering Plastics,* Vol 2, *Engineered Materials Handbook,* ASM International, 1988

CHAPTER **4**

Grinding and Polishing

GRINDING removes saw marks and levels and cleans the specimen surface. Polishing removes the artifacts of grinding but very little stock. Grinding uses fixed abrasives—the abrasive particles are bonded to the paper or platen—for fast stock removal. Polishing uses free abrasives on a cloth; that is, the abrasive particles are suspended in a lubricant and can roll or slide across the cloth and specimen. A book edited by Marinescu et al. describes in detail the scientific aspects of ceramic abrasion processes.

Some companies do not distinguish between grinding and polishing, as in the previous paragraph, but use the term *lapping* to mean grinding or coarse polishing with an abrasive slurry against a hard metal platen. Lapping is used in ceramography and ceramic manufacturing to produce extremely flat surfaces.

Ceramographic specimens can be ground and polished manually, but automatic machines usually yield better-quality, faster, more reproducible results. Manual grinding allows better control of grinding depth than automatic grinding, which could be important when the cross section at a specific depth is of interest. Automatic equipment is much more expensive than manual machines.

Diamond abrasives are recommended for grinding most ceramics, but silicon carbide (SiC) paper and cubic boron nitride (CBN) platens can also be used. End each abrasive step when the artifacts (e.g., cracks or scratches) imparted by the previous step are completely removed. Grit sizes of abrasives and micron sizes are correlated in Appendix C.

Reference 1 provides a number of material-specific automatic grinding and polishing methods.

Automatic Grinding

The pressure, time, and starting abrasive size depend on the number of mounts being ground, the abrasion resistance of the ceramic, the amount of wear on the abrasive particles, and the smoothness of the as-sawed surface. An automatic grinding and polishing machine is shown in Fig. 4.1.

Automatic grinding method steps are:

1. Symmetrically load three to six mounted specimens into the specimen holder of an automatic grinding-polishing machine, with the flat surface of the ceramic section downward. Most manufacturers provide a leveling tool for loading the mounts into the holder. Attach the holder to the polishing head.
2. Grind the specimens at a contact pressure of 40 to 150 kPa on a bonded diamond platen for approximately 60 s or until the exposed surface of each specimen is flat and clean. Note that the pressure indicated on the grinding machine gage is usually the incoming air pressure, which is not necessarily equal to the pressure of the specimens against the platen. Perforated or grooved platens are available that aid in the removal of swarf. Experiment with the abrasive size, contact pressure, relative rotation directions (same or opposite), and frequencies shown subsequently to attain the best results. Typical machine settings:

Abrasive	240–400-grit (60–40 μm) metal- or resin-bonded diamond
Time	30–60 s (or until specimen is flat and saw marks are removed)
Lubricant	Water sprayed continuously
Head air pressure (e.g., Buehler machines)	210 kPa, 2.1 bar, or 30 psi
Head force (e.g., Struers machines)	200 N
Platen frequency	200–300 rpm
Head frequency	100–150 rpm
Head direction	Opposite to platen

   Contact pressure or frequency that is too high could damage the specimens or machine and shorten the life of the polishing cloth in the polishing steps. Contact pressure or frequency that is too low slows the rate of stock removal and can prevent any significant abrasion at all.
3. Remove the specimen holder from the machine and clean the specimens, as in Subroutine 4.1, but do not remove the specimens from the holder until the last polishing step is complete. Once clean, return the specimen holder to the machine for polishing or more grinding in successive steps on ever-finer abrasives and follow each step with thorough cleaning. In many cases, all the grinding can be accomplished in a single step, such as in the procedure described in Table 4.1.

Automatic Polishing

After the finest grinding step, polish the specimens on napless polishing cloths loaded with lubricant and progressively smaller diamond abrasives.

Subroutine 4.1: Cleaning Ceramographic Mounts

After each abrasive step, rinse each specimen in warm tap water. Do not remove specimens from the holder if an automatic polishing machine is being used. Use distilled or deionized water if the tap water is too hard. Keep a 250 to 400 mL beaker of distilled water containing a laboratory detergent, such as Micro-90 or Alconox, on a hotplate at 60 to 80 °C. Swab each specimen with a cotton ball soaked in the warm, soapy water. If the ultrasonic bath is wide enough, ultrasonicate the entire specimen holder. Support the holder on a ring mold or something similar to prevent the polished faces from touching the basket or tank. Rinse each specimen again in warm tap water or deionized water.

After the last abrasive step, remove the mounts from the holder if automatically polished, swab as mentioned previously, and clean one more time in warm distilled water in an ultrasonic bath for 1 to 2 min. Quickly remove each mount from the ultrasonic bath, rinse with distilled water, spray with ethanol, and dry under a heat gun.

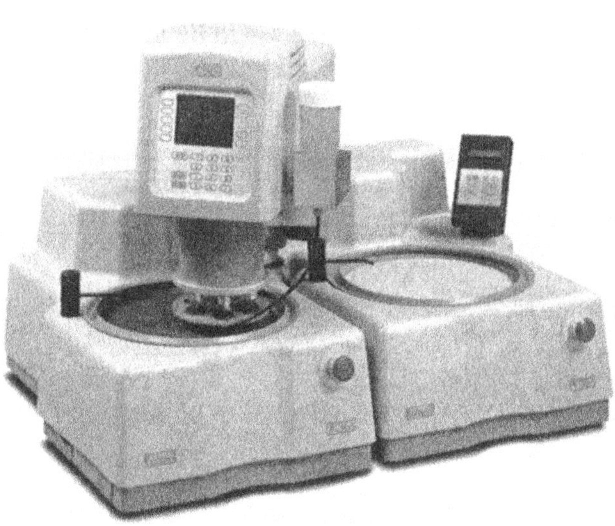

Fig. 4.1 Automatic grinding and polishing machine

Diamond polishing abrasives are typically available in 30, 15, 9, 6, 3, 1, and 0.25 μm sizes, in liquid suspensions, pastes, and aerosols. The suspensions can be automatically sprayed by some machines at timed intervals. Not every diamond size available is used or necessary in the procedure in Table 4.1. The transition from grinding to polishing may require additional time on the coarse polishing step to remove the artifacts of grinding. If paste is used, reapply it to the polishing cloth every few minutes. All types of diamond abrasives break down quickly and should be replenished frequently. Follow each polishing step with a thorough cleaning, as in Subroutine 4.1.

Use napless cloth for diamond pastes or suspensions and napped cloth for the alumina slurry or colloidal silica. Napless cloth is a stiff, nonwoven PVC chemotextile sold under such trade names as Texmet, Pellon, DP-Plan, MD-Plan, and Pan-W. Nonwoven, fiber-reinforced-resin perforated pads and woven silk also work well for polishing ceramics with diamond pastes and suspensions. Flocked twill or napped cloth has a fuzzy texture that conforms to the surface being polished.

Spread diamond paste, if used, on the cloth with a clean, gloved finger, along with additional lubricant. Polishing lubricants come under various names, including lapping oil, diamond extender, and blue lubricant. Be careful not to contaminate the paper with larger-size abrasive particles. Replace torn cloths immediately, being careful to smooth out any wrinkles or bubbles in the new cloth. Use xylene to dissolve the adhesive when removing worn-out cloth from the platen. Wear rubber gloves when using xylene. A worn-out cloth is easier to remove if the platen is first warmed with a heat gun. Platens tend to heat up during polishing and may require air cooling between intervals in order to prevent the polishing cloth from peeling or rupturing.

Step 5(a) in Table 4.1, relief polishing, is optional. Relief polishing is not recommended when the specimen is to be tested for microhardness;

Table 4.1 A typical ceramographic grinding and polishing procedure for an automatic polishing machine

Step	Abrasive and lubricant	Time, min	Platen frequency, rpm	Head frequency, rpm
1. Plane grinding	240-grit bonded diamond disc sprayed continuously with water	0.5–1 (or until specimen is flat and saw marks are removed)	200–300	120–150 opposite to platen
2. Coarse polishing	15 μm diamond suspended in water-soluble oil, sprayed every 20–30 s on napless paper	5–10	120–150	120–150 opposite to platen
3. Polishing	6 μm diamond suspended in water-soluble oil, sprayed every 20–30 s on napless paper	5–10	120–150	120–150 opposite to platen
4. Fine polishing	1 μm diamond suspended in water-soluble oil, sprayed every 20–30 s on napless paper	5–10	120–150	120–150 opposite to platen
5(a) Relief polishing (optional)	0.05 μm γ-Al_2O_3 slurry sprayed every 20–30 s on napped cloth	1–5	120–150	120–150
5(b) Vibratory polishing (optional)	Colloidal silica slurry, replenished every 30–60 min on napped cloth	60–480	...	...

Note: For machines without timed spraying, the slurries can be poured from squeeze bottles or aerosols, or diamond pastes can be used instead.
Source: Ref 2

when edge retention is critical, such as on thin plates; or when the specimen will be viewed in high magnification, such as fine-grained microstructures. Relief polishing in conjuction with Nomarski differential interference contrast (see Chapter 7) can enhance the contrast at low magnification by means of differential abrasion rates between harder and softer phases, for example, Al_2O_3 and intergranular glass in 85 to 98% alumina compositions, SiC and silicon in reaction-bonded silicon carbide, and between adjacent grains of $MgAl_2O_4$ spinel. Relief polishing can also polish the metal components in cross sections of microelectronic devices.

Vibratory polishing with colloidal silica or alumina slurry, step 5(b) in Table 4.1, is another final polish technique. Each mounted specimen is clamped into a heavy brass or stainless steel cuplike holder. The weighted mount glides freely around a damp, napped polishing cloth on a vibrating platen for hours at a time. This method works very well for soft metals and semiconductors and is useful for some harder metals and ceramics. Ceramics that have low abrasion resistance and are not easily polished, such as AlN oftentimes, may be adequately polished by vibration on colloidal silica for 8 h. The colloidal silica suspension should be replenished every hour or so, a few milliliters at a time, and the napped cloth must remain damp.

In some cases, a corrosive liquid is used along with the relief polishing slurry in a technique called *attack polish.* Attack polish combines mild etching and final polishing into a single step. Colloidal silica is suspended in a caustic solution that has an attack-polish effect on some materials. Either colloidal silica or a 1 to 10 mixture of Murakami's solution (see Table 5.1) to 0.05 μm γ-Al_2O_3 is recommended for the final polishing step on alumina with an abundant glass phase (Ref. 1). Murakami's solution is 3 g KOH and 30 g $K_3Fe(CN)_6$ in 60 mL distilled water. The attack polishing slurry is applied to chemically resistant synthetic fiber cloth rotating at 120 rpm for 30 min. The load is 15 N per 31.8 mm (1.25 in.) mounted specimen.

Manual Grinding

The manual method is useful when automatic equipment is not available or when the depth of grinding is critical. Cross sections of microelectronic devices, such as multiplayer packages, often must be ground to a specific depth.

To grind a ceramographic section manually, choose a reference point on the specimen, such as point *Q* in the 12 o'clock position shown in Fig. 4.2(a). Hold the specimen surface firmly against the abrasive disc or belt such that the reference point is fixed with respect to the direction of abrasive motion. Continue grinding until the saw marks are replaced by the parallel scratches of the first abrasive, as in Fig. 4.2(b). Clean the ground surface as described in Subroutine 4.1. Rotate the reference point *Q* to the 3 o'clock position, as in Fig. 4.2(c), and grind the specimen on the next finer abrasive until the previous artifacts are removed.

The new parallel scratches lie at a 90° angle to the previous ones, as in Fig. 4.2(d). Rotation of the mount by 90° after each abrasive step (Fig. 4.2e) allows one to easily see when the artifacts of the previous preparation step have been removed. Clean the mount thoroughly after each step, as in Subroutine 4.1, to prevent transfer of abrasive particles from one platen to the next.

In many cases, all the grinding can be accomplished in a single step.

Manual Polishing

After the finest grinding step and subsequent cleaning, manually polish the specimen on napless polishing cloths loaded with lubricant and 15, 6, and 1 μm diamond paste, respectively. Rotate the specimen 90°, as in Fig. 4.2(a–e), and clean it thoroughly, as in Subroutine 4.1, after each polishing step. The relief polishing step with 0.05 μm γ-Al_2O_3 suspension is optional. The evolution of the microstructure as a result of each polishing step is demonstrated in Fig. 4.3(a–d).

Grinding and Polishing Accessories

Lubricants. Grinding and polishing lubricants are widely used in ceramography; dry grinding is extremely rare. The lubricant facilitates the interaction between the abrasive and the specimen, whether the abrasive is

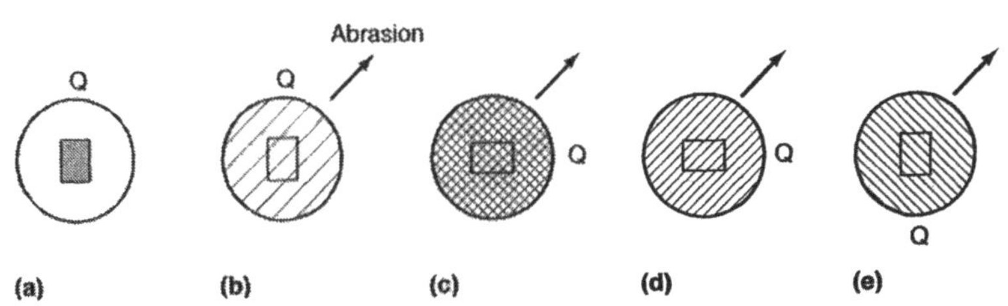

Fig. 4.2 Sequence of steps in manual grinding and polishing (Ref 2). (a) As-sawed, as-mounted surface. (b) The surface in (a) has been removed by a coarse abrasive. Point *Q* is fixed with respect to the abrasive direction indicated by the arrow. (c) The mount has been rotated 90°, and the surface in (b) has been partially removed by the next finer abrasive. (d) The finer abrasive has removed all the artifacts from the previous abrasive step. (e) For the next step, *Q* is rotated to the 6 o'clock position and ground or polished until this surface is removed by the next finer abrasive. Point *Q* would be rotated again, to the 9 o'clock position, for the step after that.

fixed or free. The grinding lubricant acts as a coolant to prevent heat buildup from friction, transports the swarf away from the platen and specimen, and uniformly distributes the contact stresses between the platen and the specimen during grinding. The polishing lubricant adheres the abrasive and swarf to the paper to prevent dust, enables the abrasive

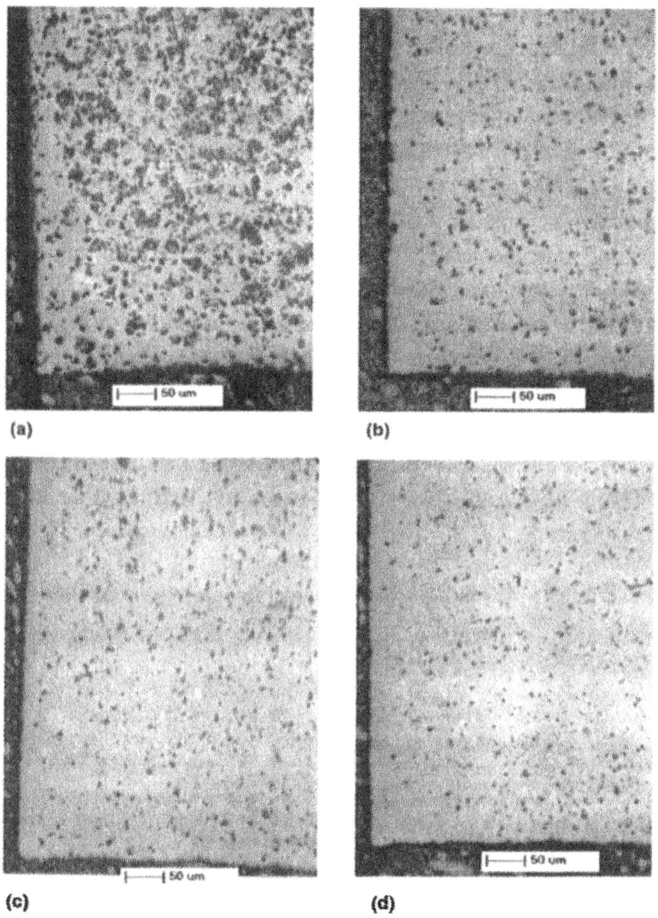

Fig. 4.3 Evolution of microstructure in a 96% Al_2O_3 substrate after a series of manual polishing steps. (a) Substrate after manual polish with 15 μm diamond paste. The top and left edges of the photo are also the specimen edges. The concentration of voids increases from edge to center and therefore are not true pores. The scratches from the abrasive are horizontal. (b) The same corner of the substrate after manual polish with 6 μm diamond paste. The scratches from the abrasive are now vertical, the specimen having been rotated 90°. The apparent "porosity" has decreased significantly. (c) The same corner of the substrate after manual polish with 3 μm diamond paste. The scratches are horizontal once again, following another 90° rotation of the specimen. The size and quantity of pluckouts continue to decrease. (d) The same corner of the substrate after manual polish with 1 μm diamond paste. The specimen was initially rotated 90° and polished linearly to remove the 3 μm scratches, then polished in a rotational motion opposite to the polishing wheel such that the last scratches are in all directions. The dark spots in the photo are mostly true pores rather than pluckouts.

particles to roll and slide easily between the paper and the specimen, and uniformly distributes the contact stresses between the paper and the specimen during polishing.

Tap water is the most common lubricant for grinding and lapping. The water can be recirculated from a reservoir, but the reservoir should contain a series of weirs and an outlet filter to trap the swarf and prevent it from being recirculated along with the water. Corrosion-inhibiting chemicals and algicides can be added to the reservoir. The water in the reservoir should be replenished often, even daily. The reservoir should be thoroughly cleaned and filled with fresh water every few months, depending on frequency of use. Specimens that are soluble in or easily corroded by water should be ground with a nonaqueous lubricant, such as oil or filtered kerosene.

Water-soluble lubricants that have an oily feel are recommended for most polishing applications. The water-soluble lubricants are easily washed away in warm tap water or by a swab in warm, soapy water. Oil-based and other nonaqueous lubricants can be used for polishing but are not as easy to remove. Nonaqueous lubricants are recommended for ceramics and minerals that are easily dissolved in water. Diamond pastes and slurries are usually available as either water- or oil-based suspensions. Diamond pastes and aerosols require additional lubricant on the polishing paper. Polishing lubricants are sold under names such as blue lubricant or diamond extender.

Abrasives. While diamond particles are emphasized in this text as the abrasive of choice for ceramographic grinding and polishing, they are not

Table 4.2 Polishing textiles and their characteristics

Cloth	Trade names	Description and applications
Billiard	. . .	Wool sheared pile for rough polishing of ferrous metals with alumina or diamond
Canvas	Duck cloth	Rough polishing of metals
Cotton	Metcloth, MD/DP-Mol	Tightly woven cotton for rough polishing of metals with alumina or diamond
Flocked twill	Lecloth, Microcloth, SP-PoliFloc, Suede Cloth, DP/MD-Plus	Rayon fibers woven in a cotton back; medium nap; for final polishing with gamma alumina
Metal mesh	Ultra-Plan, DP-Net	Wire cloth for coarse polishing harder materials
Napless	Texmet, Pellon, DP-Plan, MD-Plan, Pan-W	Nonwoven PVC chemotextile for coarse and fine polishing with diamond or alumina
Nylon	Imperial, DP-Nylon	Napless nylon for intermediate polishing with diamond
Perforated pad	Polimet	Hard, nonwoven, fiber-reinforced resin with an array of perforations for fast removal of hard materials
Felt	Red felt, OP-Felt	Wool plucked pile for intermediate polishing of ferrous metals with diamond
Silk	Technotron, DP/MD-Dur	Woven silk; hard cloth; for intermediate polishing with diamond. Excellent for keeping hard materials flat
Selvyt	Velveteen, MD/DP/OP-Nap	Medium nap cotton for intermediate polishing with diamond or alumina
Velvet	Rayvel, Mastertex	Synthetic velvet with long nap; for final polishing of soft materials

Source: Ref 4

the only abrasives on the market. Diamonds are available in both natural (mined) and synthetic (made in a factory) forms. The synthetic diamonds can be manufactured to have better cutting edges than the natural diamonds.

Silicon carbide paper is widely used in metallographic grinding and can be used for ceramics, too, but tends to be short-lived. Cubic boron nitride is available in metal-bonded discs, and CBN discs and 240-grit SiC paper can be used for coarse polishing in lieu of metal-bonded diamond discs. Reference 3 describes the characteristics of abrasives and abrasion of ceramics.

Powders of silicon carbide, cerium oxide, cesium oxide, ferric oxide (jeweler's rouge, Fe_2O_3), gamma alumina, chromia (Cr_2O_3), magnesia (MgO), and colloidal silica are used in some metallographic and lapidary polishing applications and may have niche applications in ceramography. Colloidal silica is submicron crystalline SiO_2 particles suspended in a caustic solution and is commonly used with vibratory polishers to final-polish AlN and many other ceramics. Gamma alumina (γ-Al_2O_3), formed by incomplete dehydroxylation of $Al(OH)_3$, is metastable in the cubic phase and used for relief polish. In general, the abrasive should be harder than the specimen, which is why diamond is preferred for grinding and polishing ceramics.

Some abrasive discs have channels, perforations, or gaps between clusters of abrasive particles to fa cilitate swarf removal. The perforated discs are preferred by many ceramographers over conventional fixed-abrasive discs.

Polishing Textiles. Most of the metallographic consumables vendors listed in Appendix B can provide samples of the many kinds of polishing cloths that they offer, on request. The samples typically come as 2 cm circles attached to a paperboard display, with a brief description of each type. Only the nonwoven, napless type of cloth is recommended in this text for polishing with diamond abrasives, although other cloths may have applications in ceramography, especially the perforated pads. The napped, flocked twill is recommended only for relief polishing with colloidal silica or γ-Al_2O_3. Some other polishing cloths are described in Table 4.2 (Ref 4). Magnified images of the textile fibers are shown in Ref 5.

REFERENCES

1. G. Elssner, H. Hoven, G. Kiessler, and P. Wellner, *Ceramics and Ceramic Composites: Materialographic Preparation,* R. Wert, Trans., Elsevier Science Inc., 1999, p 74–133
2. R.E. Chinn, Preparation of Microstructures of Alumina Ceramics, *Structure,* Vol 33, 1998, p 16–20
3. E. Ratterman and R. Cassidy, Abrasives, *Ceramics and Glasses,* Vol 4, *Engineered Materials Handbook,* ASM International, 1991, p 329–335

4. L. Dillinger, "Polishing," Met-Tips 13, Leco Corp., 1985
5. G.F. Vander Voort, *Metallography: Principles and Practice,* ASM International, 1999, p 104–107

SUGGESTED READING

- I.D. Marinescu, H.K. Tonshoff, and I. Inasaki, Ed., *Handbook of Ceramic Grinding and Polishing,* Noyes Publications/William Andrew Publishing, LLC, 2000

CHAPTER 5

Etching

ETCHING usually implies some sort of selective corrosion for enhancement of the microstructure. In the context of ceramography, etching includes several noncorrosive techniques. Etching reveals and delineates grain boundaries and other microstructural features that are not apparent on the as-polished surface. The delineated grain boundaries can be used for grain size measurement or comparison. Do not etch the microstructure before measuring the microindentation hardness or evaluating the porosity. Etching is generally not applied to thin sections.

Etching types are classified as electrochemical, physical, and optical (Ref 1). These classifications are arbitrary and can be considered in other ways. For practical reasons, etching types are classified here as aqueous chemical, thermal, and other. The "other" category broadly includes relief polish, plasma, electrolysis, ion or cathodic vacuum etching, molten salt immersion, heat tinting, and interference-layer types of etching. The optical techniques discussed in Chapter 7 can be used in conjunction with the thermal, chemical, and other methods.

Etching, perhaps more than any other aspect of ceramography, is short on reproducibility and depends heavily on the experience of the technician. Each method described in this chapter might be very effective in one laboratory and much less effective on the same ceramic in another lab. Most of the etching methods depend to some degree on the subtleties of the process, including temperature, time, pressure, concentration, atmosphere, voltage, current, secondary phases, solutes, porosity, grain size and orientation, fabrication process, grinding and polishing techniques, and many other process variables. Time is usually the easiest parameter to reduce or extend in order to vary the degree of etching, although other parameters, such as temperature and concentration, are often more effective.

Some etching procedures are given in Table 5.1 and in Ref 1–5.

Thermal Etching

While thermal etching is not etching in the sense of selective corrosion, it is often the simplest, cleanest, and most effective method of delineating

Table 5.1 Summary of etching techniques for various ceramics

Ceramic	Thermal etch	Chemical etch	Other etching methods
Alumina, α-Al_2O_3	T = 1500 °C t = 20–30 min Atm = air or inert gas Prep = Subroutine 5.1 (Ref 1) T = 1100–1300 °C t = 2 h Atm = air Prep = Subroutine 5.1	Reagent = concentrated $H_3PO_{4(aq)}$ T = boiling t = 15–60 s Prep = Subroutine 5.1 (Ref 5) Reagent = concentrated $H_2SO_{4(aq)}$ T = boiling t = 2–10 min Prep = Subroutine 5.1	• Relief polish with 0.05 μm Al_2O_3 • Petrographic thin section in polarized transmitted light • Molten salt etch in V_2O_5 for 60 s at 750–900 °C (Ref 6) • Ion etch at 6 kV and 30μA in vacuum for 45 min (Ref 2) • Thermochemical etch 2–3 min in a 3:1 mixture of 0.8N H_3PO_4 and 0.5N HNO_3, then 10 min at 1150 °C (Ref 7)
Alumina, β-Al_2O_3	(Ref 5) T = 1470 °C (200 K below the sintering temp) t = 1–5 min Atm = air Prep = Subroutine 5.1	...	...
Aluminum nitride, AlN	...	Reagent = equal volumes glacial acetic acid and $HNO_{3(aq)}$ T = boiling t = 15–30 min Prep = Subroutine 5.1	• Relief polish with 0.05 μm Al_2O_3 • Thermochemical etch 2–3 min in a 2:1 mixture of 0.8N H_3PO_4 and 0.5N HNO_3, then 10 min at 1100 °C (Ref 7).
Barium titanate, $BaTiO_3$	...	Reagent = zirconia etchant (see ZrO_2) T = boiling t = 15–60 s (Ref 5) Reagent = 1:3 $HF_{(aq)}$ + concentrated $HCl_{(aq)}$ T = room temperature t = 7 min–2 h	...
Beryllia, BeO	(Ref 1) T = 1200 °C t = 3–10 min Atm = H_2 at P = 500 torr Prep = Subroutine 5.1	(Ref 1) Reagent = 40% $HF_{(aq)}$ T = room temperature t = 2 s–6 min	...
Boron carbide, B_4C	...	Reagent = $NaOH_{(aq)}$ T = boiling t = 1–10 min Prep = Subroutine 5.1 Electrolytic (Ref 4) Reagent = 1% $KOH_{(aq)}$ Voltage = 5 V_{DC} T = room temperature t = 15 s	...
Boron nitride, BN	...	...	• Molten salt etch in 1:9 mixture of KOH KNO_3 for 1 min (Ref 2) • Optical etch with polarized light (Ref 4)
Calcia, CaO	...	(Ref 5) Reagent = conc $HCl_{(aq)}$ T = room temperature t = 35–60 min	...
Cemented carbide, e.g., WC + Co	(Ref 4) T = 400 °C t = 30–60 min Atm = unspecified Prep = Subroutine 5.1	Reagent = Murakami's etchant (see α-SiC) T = room temperature t = 10 s	• Relief polish with 0.05 μm Al_2O_3 • Interference layer, vapor deposition of ZnSe or ZnTe, or reactively sputtered Fe in O_2 (Ref 4)
Ceria, CeO_2	...	(Ref 4) Reagent = 50 mL H_2O + 45 mL (65%) $HNO_{3(aq)}$ + 5 mL (40%) HF T = room temperature t = 10–20 min	Molten salt etch in $KHSO_4$ in a Pt crucible for 3–10 min (Ref 1)
Chromia, Cr_2O_3	...	...	Molten salt etch in $KHSO_4$ in a Pt crucible for 3–10 min (Ref 1)

(continued)

Table 5.1 (continued)

Ceramic	Thermal etch	Chemical etch	Other etching methods
Cordierite, $2MgO \cdot 2Al_2O_3 \cdot 5SiO_2$	T = 1300 °C t = 10–20 min Atm = air or inert gas Prep = Subroutine 5.1	Reagent = 100 mL methanol + 3 mL $HCl_{(aq)}$ + 0.5mL $HF_{(aq)}$ T = room temperature t = 30–60 min	...
Graphite, C	...	...	• Optical etch; view in polarized light • Plasma etch in O_2 for 20–60 min at 200 W (Ref 4)
Magnesia, MgO	...	(Ref 5) Reagent = concentrated $HCl_{(aq)}$ T = room temperature t = 35–60 min (Ref 2) Reagent = concentrated $H_3PO_{4(aq)}$ T = room temperature t = 4 min	Relief polish with 0.05 µm Al_2O_3
Mullite, $3Al_2O_3 \cdot 2SiO_2$ or $Al_6Si_2O_{10}$	...	(Ref 3) Reagent = concentrated $H_3PO_{4(aq)}$ T = boiling t = 6 min Prep = Subroutine 5.1	Petrographic thin section in polarized transmitted light
Sialon, $Si_3N_4 \cdot Al_2O_3$	...	(Ref 2) Reagent = 20% $NH_4HF_{2(aq)}$ T = 82 °C t = 3–10 min Prep = Subroutine 5.1	• Ion etch at 7.5 kV and 50 µA in vacuum for 5 h (Ref 2) • Molten salt etch in 1:9 mixture of KOH + KNO_3 for 3 min (Ref 3) • Plasma etch in CF_4 + 5% O_2 for 30–45 s at 400 W and 13.56 MHz (Ref 8)
Silicon carbide, α-SiC—direct-sintered	(Ref 5) T = 1300–1500 °C t = 1–3 h Atm = vacuum Prep = Subroutine 5.1	Reagent = Murakami's etchant: 10g $K_3Fe(CN)_6$ + 10 g KOH or NaOH + 100 mL H_2O T = boiling t = 60–120 s Prep = Subroutine 5.1	• Plasma etch in $CF_4 + O_2$ (Ref 8) • Molten salt etch in $NaHCO_3$, $KHCO_3$, or $Na_2B_4O_7$ for ~ 10 min (Ref 1)
Silicon carbide, β-SiC	...	...	Molten salt etch in mixture of KOH + KNO_3 at 480 °C for 5 min (Ref 2)
Silicon carbide, SiC—reaction-bonded (contains free Si)	...	Electrolytic *Reagent* = 2 g KOH + 100 mL H_2O *Current density* = 1 A/cm² *Voltage* = 6 V_{DC} T = room temperature t = 20–180 s	• Relief polish with 0.05 µm Al_2O_3 • Ion etch at 6 kV and 30 µA for 45 min (Ref 2) • Molten salt etch in KOH for 5 min (Ref 2) • Interference layer, reactive-sputtered Fe cathode in O_2 (Ref 4)
Silicon nitride, Si_3N_4	(both Ref 5) T = 1250 °C t = 15 min Atm = vacuum Prep = Subroutine 5.1 T = 1600 °C t = 5 h Atm = N_2 Prep = Subroutine 5.1	*Reagent* = zirconia etchant + 5 mL $H_2SO_{4(aq)}$ T = boiling t = 15–30 s *Prep* = Subroutine 5.1 (Ref 1) *Reagent* = $H_3PO_{4(aq)}$ T = boiling t = 5–15 min *Prep* = Subroutine 5.1	• Plasma etch in $CF_4 + O_2$ (Ref 8) • Molten salt etch in NaOH at 400–450 °C for 1–10 min (Ref 5) • Molten salt etch in KOH for 1 min (Ref 2)
Sillimanite, $Al_2O_3 \cdot SiO_2$ or Al_2SiO_5	...	...	• Relief polish with 0.05 µm Al_2O_3 • Molten salt etch in KHF_2 for 5–10 min, rinse in $H_2PO_{3(aq)}$ (Ref 1)
Spinel, $MgO \cdot Al_2O_3$ or $MgAl_2O_4$	T = 1500 °C t = 20–30 min Atm = air or inert gas Prep = Subroutine 5.1	*Reagent* = $H_3PO_{4(aq)}$ T = boiling t = 15–60 s *Prep* = Subroutine 5.1	Relief polish with 0.05 µm Al_2O_3
Strontium-celsian, monoclinic $Sr_2Al_2Si_2O_8$	...	(Ref 9) *Reagent* = 1% $HF_{(aq)}$ + 2% $HNO_{3(aq)}$ T = room temperature t = 20 s	...
Superconductor, $YBa_2Cu_3O_7$	...	*Reagent* = glacial acetic acid T = room temperature t = 30 s	Optical etch; view in polarized reflected light

(continued)

Table 5.1 (continued)

Ceramic	Thermal etch	Chemical etch	Other etching methods
Thoria, ThO_2	...	(Ref 3) Reagent = $H_3PO_{4(aq)}$ T = boiling t = 5 min Prep = Subroutine 5.1 (Ref 1) Reagent = $H_2SO_{4(aq)}$ T = 60 °C t = 1–2 min Prep = Subroutine 5.1	...
Titanium diboride, TiB_2	...	Reagent = $H_2SO_{4(aq)}$ T = boiling t = 30–60 s Prep = Subroutine 5.1 (Ref 1) Reagent = 100 mL HCl + 10 mL $HNO_{3(aq)}$ T = room temperature t = 15 s	Relief polish with 0.05 µm Al_2O_3
Titanium carbide, TiC	Heat tint T = Bunsen flame t = 5 min or until color change Atm = air Prep = Subroutine 5.1	Electrolytic Reagent = 80 mL 8% NaOH + 80 mL H_3PO_4 + 10 mL H_2SO_4 + 10 g $CuSO_4{\cdot}5H_2O$ Current density = 0.9 A/cm^2 Voltage = 3.5 V_{DC} T = room temperature t = 30 s Copper cathode (Ref 1)	...
Titania, TiO_2	(Ref 5) T = 1350 °C t = 1 h Atm = air Prep = Subroutine 5.1	...	...
Urania, UO_2	(both Ref 1) T = 1200 °C t = 3–10 min Atm = H_2 at P = 500 torr Prep = Subroutine 5.1 T = 1600 °C t = 60 min Atm = vacuum at P = 10 torr Prep = Subroutine 5.1	(Ref 1) Reagent = $H_2SO_{4(aq)}$ T = 60 °C t = 1–2 min Prep = Subroutine 5.1	Ion etch at 4 kV and 2 mA in Kr at 40 mtorr for 10–30 min (Ref 10)
Zinc oxide, ZnO	...	(Ref 1) Reagent = 5 mL glacial acetic acid + 100 mL H_2O T = room temperature t = 30 s (Ref 4) Reagent = 5% citric acid T = room temperature t = 30 s–4 min	...

(continued)

Table 5.1 (continued)

Ceramic	Thermal etch	Chemical etch	Other etching methods
Zirconia, ZrO_2	T = 1475–1500 °C t = 10–20 min Atm = air or inert gas Prep = Subroutine 5.1 For fully stabilized cubic ZrO_2 (CSZ or ZDY) only. Not recommended for TZP, PSZ, or TTZ	Reagent = 8 g NH_4HF_2 + 8 mL $HF_{(aq)}$ + 100 mL H_2O + 1 mL $H_2O_{2(aq)}$ T = boiling t = 30–60 s After etching, remove residue with light relief polish, then soak overnight in water. (Ref 4) Reagent = 90 mL H_2O + 90 mL HNO_3 + 18 mL $HF_{(aq)}$ T = room temperature t = unspecified	• Relief polish with 0.05 μm Al_2O_3 • Ion etch at 5 kV and 50 μA in argon for 75 min at 35° incidence angle (Ref 11) • Retrographic thin section of TZP or PSZ in transmitted light
Zirconium diboride, ZrB_2	...	(Ref 3) Reagent = equal volumes concentrated HNO_3+ $H_2O_{2(aq)}$ T = room temperature t = 55 s	...

Note: Observe the cautions in Chapter 1—hydrofluoric acid ($HF_{(aq)}$) is especially hazardous to human tissue. T, temperature; t, time; atm, atmosphere; P, pressure; Prep, preparation; TZP, tetragonal zirconia polycrystals; PSZ, partially stabilized zirconia; TTZ, transformation-toughened zirconia. Source: Ref 12

grain boundaries in ceramic microstructures, especially in single-phase oxides. The flatness of a polished surface is thermodynamically unstable at high temperatures, as are the grain-boundary regions, such that each polished grain lowers its surface energy by rounding outward slightly as atoms diffuse from the grain boundaries. Two added advantages of thermal etching are that shallow scratches are "healed," and organics in the pores, such as lapping oil, are burned away. The danger in thermal etching is that smaller grains are absorbed into larger grains, given sufficient temperature and time, resulting in grain growth that changes the microstructure. Thermal etching is usually done at temperatures approximately 100 to 200 K below the sintering temperature, with time at the etching temperature limited to approximately 1 h or less. Some etching formulas call for a much longer time—say, several hours—at a lower temperature. Thermal etching is sometimes called annealing.

Thermal etching works best with single-phase oxides, such as 99.5% alumina. Multiphase oxides, including many types of porcelain, can undergo incongruent melting or reactions between dissimilar phases at high temperatures that change the microstructure. Nonoxide ceramics, in general, should be thermally etched in a vacuum or inert atmosphere to prevent oxidation on the polished surface. Nitride ceramics, for example, AlN and Si_3N_4, can often be thermally etched in pure N_2, but oxides and carbides might be nitrided as nitrogen replaces the surface anions. The grain growth of 99.9% alumina during thermal etching depends on the gaseous atmosphere, and very fine-grained specimens should not be thermally etched because of the tendency for grain growth (Ref 13).

Allotropic ceramics, such as partially stabilized zirconia (ZrO_2), tend to shatter at high temperatures due to the strain associated with phase changes. High-purity zirconia undergoes a displacive transformation from the monoclinic phase (density, $\rho = 5.83$ g/cm^3) to the tetragonal phase ($\rho =$ 6.10 g/cm^3) on heating above approximately 1000 °C, and that transformation is reversed on cooling through 1000 °C, as shown on the left vertical axis in Fig.5.1(a). The two phases have different specific volumes (the reciprocal of density), such that the residual stress generated by the volume change on transformation is enough to disintegrate the specimen. Partially stabilized zirconia (PSZ) has a metastable cubic primary phase and contains a small amount of CaO, MgO, or Y_2O_3 in solid solution. Partially stabilized zirconia uses the tetragonal-to-monoclinic transformation in its secondary phase to achieve a remarkable toughening effect. Partially stabilized zirconia and tetragonal zirconia polycrystals (TZP) should not be thermally etched. Zirconia densified with approximately 13% yttria (ZDY) is fully stabilized in the cubic structure. Zirconia densified with yttria retains its cubic structure over a broad temperature and composition range as shown in Fig. 5.1(b) and can be thermally etched.

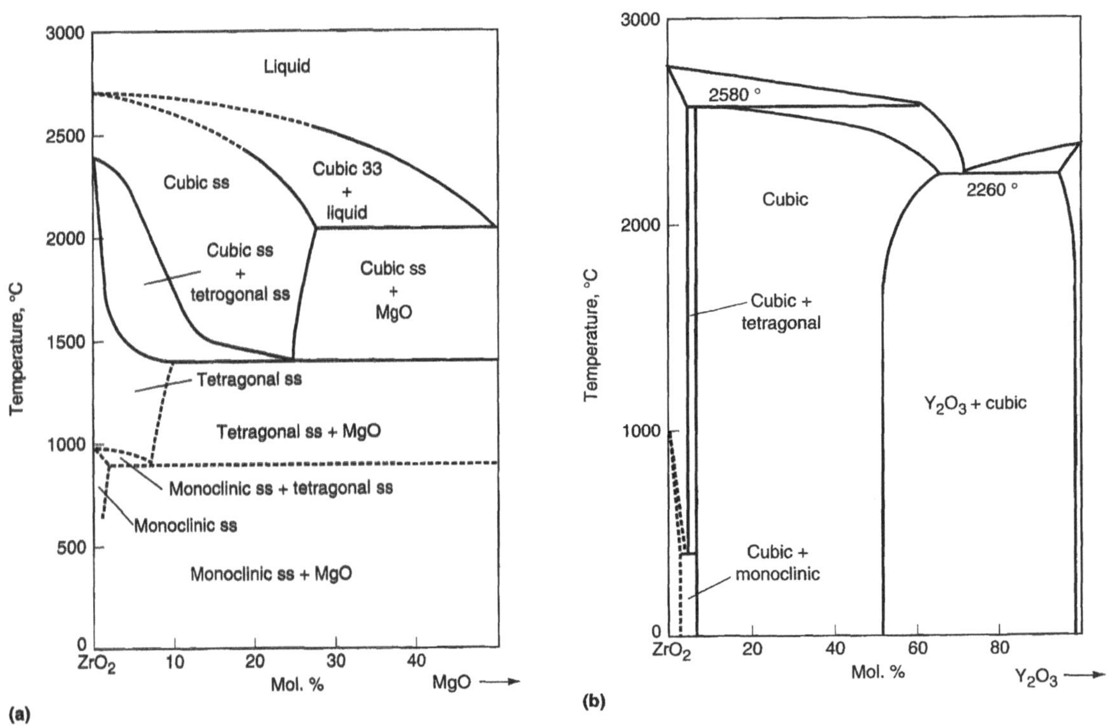

Fig. 5.1 Zirconia (ZrO_2) phase diagrams. (a) The zirconia-rich end of the ZrO_2-MgO phase diagram (Ref 14). Depending on the MgO content, the monoclinic ZrO_2 phase is thermodynamically stable up to approximately 900 °C, and the cubic ZrO_2 phase is stable as low as approximately 1400 °C. The phase diagram does not contain any information about phase transformation kinetics, but the cubic phase can be metastabilized down to room temperature with MgO in solid solution. (b) The ZrO_2-Y_2O_3 phase diagram (Ref 15). The cubic phase of ZrO_2 is stable from approximately 2300 °C down to zero when the Y_2O_3 in solid solution is 5 to 45 mol%. ss, solid solution

A general procedure for thermal etching is given subsequently, with additional details in Table 5.1. Laboratory furnaces suitable for thermal etching are available from Centorr, Deltech, Huppert, Keith, Lindberg, Thermcraft, Thermolyne, and many other manufacturers. Figures 5.2 and 5.3 demonstrate thermally etched alumina and spinel.

Thermal Etching Method. The steps are:

1. Remove the specimen from its encapsulation, as described in Subroutine 5.1.
2. A means of identification is useful in etching more than one specimen at a time. Identify each specimen, if necessary, by one or more of these methods:

   a. Weigh the specimen to four or more significant figures and record its mass.
   b. Scratch a mark on an unpolished surface with an engraver or diamond stylus.
   c. Notch one of its unpolished surfaces with a saw. The notch must be shallow to prevent fracture from thermal shock. This is best done prior to mounting.

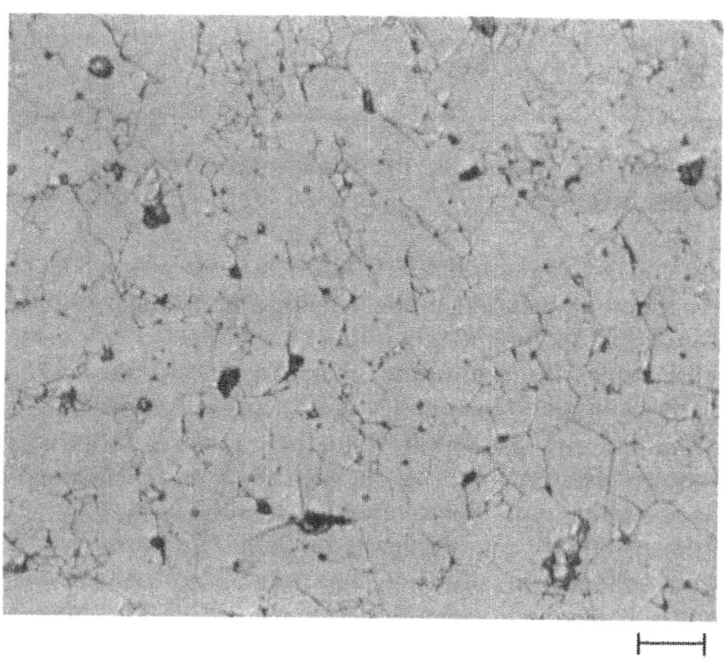

Fig. 5.2 99.5% Al_2O_3 thermally etched 30 min at 1500 °C and gold sputter coated. The mean grain size is 7.7 μm by the circle-intercept method, 8.3 μm by the lineal intersection method, and 7.2 μm by image analysis. Source: Ref 16

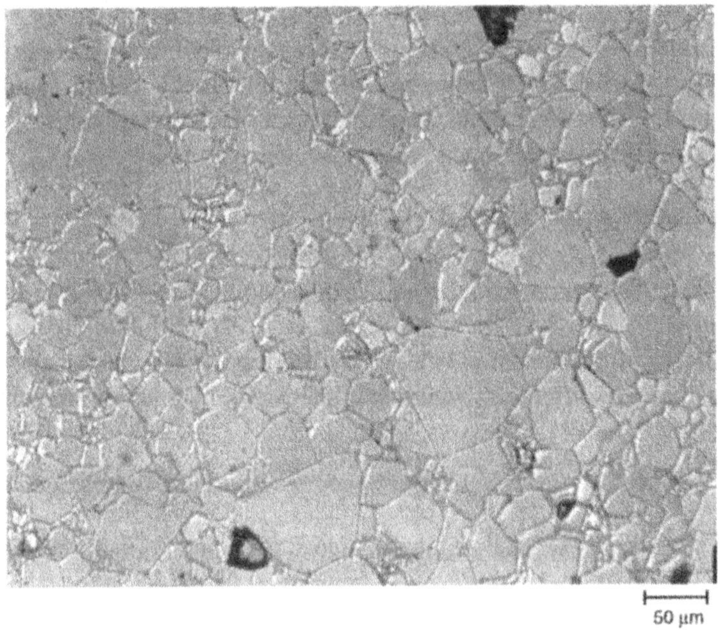

Fig. 5.3 $MgAl_2O_4$ spinel thermally etched 60 min at 1500 °C and gold sputter coated. The mean grain size is 19 μm by the circle-intercept method, 20 μm by the lineal intersection method, and 19 μm by image analysis. (Ref 16)

Subroutine 5.1: Removing the Encapsulation from a Mounted Specimen

Saw the resin encapsulation around the specimen on all sides with a diamond wafering blade, leaving a border approximately 1 mm wide. Calcine the remaining encapsulation in a ceramic crucible over a Bunsen burner flame inside a fume hood. Air cool the ash-laden specimen back to room temperature and brush off the ash. Use an ultrasonic bath, as in Subroutine 3.1, to remove the remainder of the ash residue. Spray the wet, clean specimen with ethanol and dry it with a heat gun.

If the specimen cannot tolerate the heat of a flame, try using organic solvents to dissolve the remaining encapsulation. Acrylic will slowly dissolve in warm acetone. Warm acetone is very volatile (emits vapors) and ignites easily—use it in a fume hood, away from sparks, flames, and other flammables.

3. Place the specimen in a boat or crucible of platinum or 99.8% alumina and cover it with another boat or crucible lid. Granular or tabular alumina is optional as crucible lining. *Be sure the specimen does not react with the crucible or lining at high temperatures.* Place the specimen container inside a box or tube furnace with a clean interior and clean heating elements. Close the furnace and start the purge gas flow or vacuum pump, if applicable. Heat the furnace to the etching temperature, approximately 1500 °C for alumina and fully stabilized zirconia, at a rate of 2 to 10 K/min. Maintain the furnace at the etching (soak) temperature until grains are sufficiently delineated, typically 5 to 30 min for alumina and fully stabilized zirconia. Cool the furnace back down to room temperature at 2 to 10 K/min. Smaller furnaces can heat and cool safely faster than larger furnaces. The higher rates could cause thermal shock failure of the tubes, refractory lining panels, or heating elements. The etching temperature is typically 100 to 200 K below the sintering temperature, but thermal etching can also be accomplished with lower temperatures and longer soak times. Suggested soak times and temperatures for common ceramics are listed in Table 5.1. The crucible enclosure is optional but protects the specimens from attack by contaminants from other materials that have been in the furnace, and dust. Alumina is especially susceptible to infiltration by chromium oxide, which leaves a pink stain on the white alumina.
4. If the furnace does not have a programmable temperature rate, then put the specimen container initially where it will heat most slowly, such as at the end of the tube in a tube furnace. Push the container incrementally with an alumina rod toward the zone of maximum temperature so as to minimize the possibility of thermal shock. Hold the specimens at the soak temperature for the prescribed time, then terminate the power to the furnace. Withdraw the container incrementally or allow the specimens to furnace cool.
5. When the specimen has cooled to room temperature, clean and dry it, as in Subroutine 3.1 if necessary. Reweigh each specimen to verify its identity, as needed. A fine-point felt tip pen or 0.5 mm pencil is useful for writing an identification number or letter on an unpolished surface.
6. Examine the polished surface under a microscope and repeat steps 3 to 5 until the grain boundaries are sufficiently delineated.
7. Sputter coat the polished, etched surface with a 10 to 100 nm layer of gold alloy to increase the reflectance. Sputter coating is optional for microscopic examination but is required for examination of nonconductive materials by scanning electron microscope. A gold coating gives the polished, etched surface a uniform reflectance and eliminates any subsurface reflections. Sputter coating is described in Chapter 7.

Chemical Etching

Chemical etching selectively corrodes regions of high susceptibility, such as grain boundaries, when the ceramographically polished specimen is immersed in a corrosive solution. In this context, chemical etching implies etching performed with aqueous solutions, without external electric current. Most ceramics, especially oxides, are highly resistant to corrosion. Thus, chemical etching of ceramics often requires higher concentrations of etchant solutions, higher temperatures (often boiling), and longer residence times than metals and plastics. Plastic encapsulations often corrode more easily than the ceramic specimens and can interfere with the etching process, such that the encapsulation must be removed prior to etching, in many cases. Figures 5.4 and 5.5 demonstrate chemically etched alumina and spinel.

Hydrofluoric acid ($HF_{(aq)}$)is regarded as the all-purpose, if-nothing-else-works chemical etchant for ceramics. Hydrofluoric acid rapidly attacks glass, including Pyrex beakers, and many oxide ceramics, even at room temperature. It is useful for etching many porcelains and silicates that are not listed in Table 5.1. Hydrofluoric acid is sold in plastic bottles and should be used as an etchant only in plastic beakers. Boiling etchants that contain hydrofluoric acid should be heated in a thick-walled Teflon beaker and watched carefully in case it boils dry. Teflon has a higher melting temperature than most plastics, approximately 327 °C, and is very corrosion resistant. Hydrogen fluoride gas ($HF_{(g)}$) is evolved when hydrofluoric acid evaporates or boils dry, and $HF_{(g)}$ can be as destructive to respiratory tissue as $HF_{(aq)}$ is to skin. Protective outer gear—face shield, rubber apron, rubber gloves, long pants, close-toed shoes, and so on—must always be worn by anyone handling hydrofluoric acid, and all $HF_{(aq)}$ processing should be done in a fume hood or otherwise well-ventilated room. Hydrofluoric acid burns should receive medical treatment immediately, because untreated hydrofluoric acid burns can become gangrenous.

Chemical Etching Method. The steps are:

1. Remove the encapsulation from the mounted specimen, as in Subroutine 5.1, as necessary. This step is unnecessary for some chemical etchants, such as those that are applied at room temperature.
2. Prepare the chemical etchant in a beaker and heat it on a hotplate in a fume hood, as necessary. Phosphoric acid for alumina etching must be boiled until it is cloudy and viscous. Use a Teflon beaker if the etchant attacks glass, such as etchants containing hydrofluoric acid. Carefully observe the safety practices outlined in Chapter 1 when using chemical etchants, especially with regard to MSDS, eyewear, ventilation, and protective clothing! Chemical etchants (reagents) for common ceramics are listed in Table 5.1.

Fig. 5.4 98% Al_2O_3 chemically etched 30 s in boiling phosphoric acid and viewed in differential interference contrast. The mean grain size is 20 µm by the circle-intercept method.

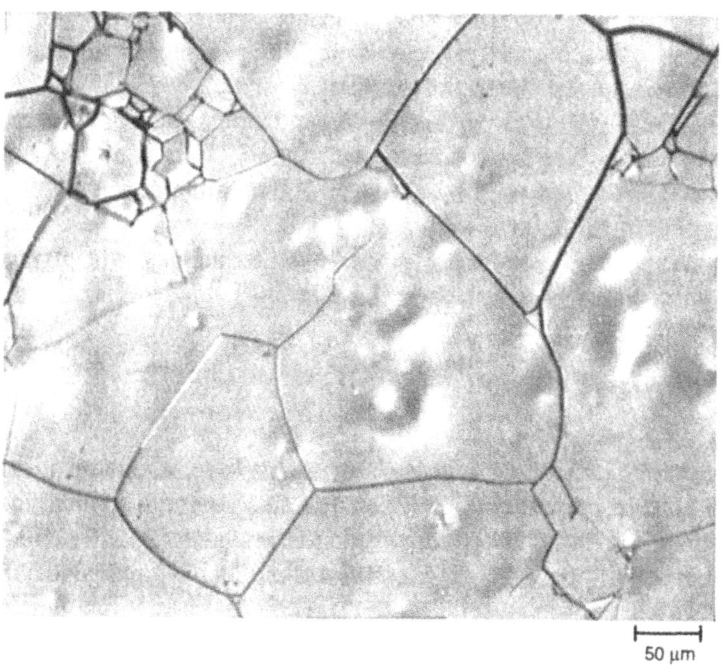

Fig. 5.5 $MgAl_2O_4$ spinel chemically etched 2 min in boiling phosphoric acid and viewed in differential interference contrast. The mean larger (~80% by volume) grain size is 160 µm, and the mean smaller (~20% by volume) grain size is 26 µm by the circle-intercept method.

3. Fill a second beaker with distilled water and warm it on the hotplate. This beaker is used to halt the etching action and rinse the etchant from the specimen. For ceramics that are especially sensitive to thermal shock, the rinse water should be at the same temperature as the etchant.
4. When the etchant is ready (e.g., boiling), immerse the specimen in the etchant and begin timing. Some ceramics etch in a matter of seconds, while others require several minutes or even hours. Apply room-temperature etchants to the polished surface with a Puritan (Hardwood Products Company) cotton-tipped applicator or equivalent, rather than by immersion. Quickly remove the specimen from the etchant with tongs and immerse it in the beaker of water to stop the etching action. Ultrasonicate the specimen in the water for 30 to 60 s and rinse and dry it as in Subroutine 3.1.
5. Examine the polished surface under a low-power microscope to look for grain boundaries. Repeat step 4 until the grain boundaries are sufficiently delineated. Several etching intervals may be required for a complete etch. Chemical etchants tend to attack scratches and pluck-outs along with grain boundaries, so further etching may result in more artifacts.
6. Some etching procedures deposit a residue on the polished surface, such as the zirconia chemical etching procedure in Table 5.1. To remove the residue, spray some 0.05 μm γ-Al_2O_3 slurry on a napped polishing cloth and rub the polished surface against the slurry for a few seconds. Ultrasonicate, rinse, and dry the specimen as in Subroutine 3.1 to remove the slurry and residue.
7. Specimens etched in hydrofluoric acid could damage microscope lenses by emitting hydrofluoric acid vapors. To prevent this, soak the specimen in a beaker of warm water or dry it in an oven at 120 °C or higher for 1 to 2 h after etching.
8. Sputter coat the specimen, if desired, to increase the reflectance. Sputter coating is described in Chapter 7.

Electrolytic Etching

Electrolytic, anodic, or *electrochemical etching* is chemical etching with an electric current applied by an external source to induce the galvanic effect in electrically conductive ceramics, such as ZrB_2, TiC, TaC, B_4C, or reaction-bonded SiC. Electrolytic etching is applicable only to electrically conductive materials, which precludes most structural ceramics. Silicon carbide is actually a semiconductor—its resistivity is much higher than metals. Electrolytic techniques are widely used in metallography for polishing as well as etching. Ceramics to be electrolytically etched can be mounted in conductive resins (see Chapter 3) that contain metal or carbon particles. Several of the equipment manufacturers listed in

Appendix B offer an electrolytic polishing and etching kit, or one can be custom-built with a direct current power supply. An electrolytic cell is shown in Fig. 5.6(a). The electrolytic process is controlled by the parameters in Table 5.2.

The TiC in Fig. 5.6(b) was electrolytically etched in 4 min at 5 V_{DC} in a 1% KOH solution (2 g KOH in 200 mL H_2O). The current density was 5 A/cm^2. The etched layer is shallow and nonuniform and can be removed with colloidal silica on a vibratory polisher in less than 1 h. A circular region approximately 1 mm in diameter surrounding the point of contact with the anode lead was not etched at all, while some of the farthest regions from the point of contact were overetched. The B_4C in Fig. 5.6(c) was etched approximately 6 min in the same solution with a current density of 1.2 A/cm^2.

Electrolytic Etching Method. The steps are:

1. The specimen usually does not need to be removed from its encapsulation, but electrical contact between the specimen and the anode lead is

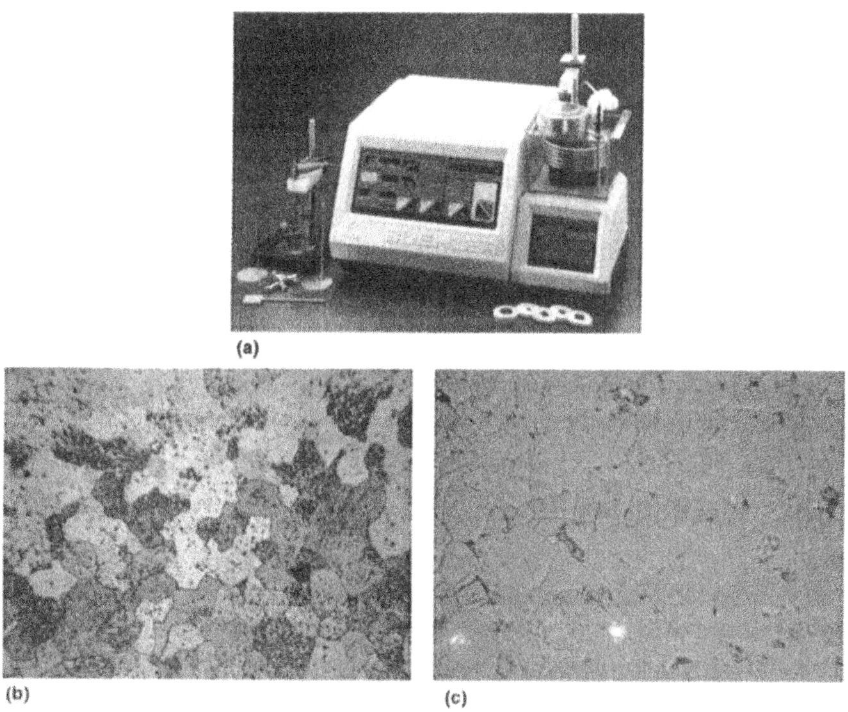

Fig. 5.6 Electrolytic etching. (a) Electrolytic cell for etching conductive specimens. (b) TiC electrolytically etched 4 min at 5 V direct current in 1% $KOH_{(aq)}$. The severity of the etch and the distance from the electrode contact point increase from the top of the image to the bottom. (c) B_4C electrolytically etched 6 min at 5 V direct current in 1% $KOH_{(aq)}$. Viewed in differential interference contrast

Table 5.2 Electrolytic etching parameters and typical values for reaction-bonded SiC

Parameter	Typical value
Current density	1 A/cm^2
Electrode composition and shape	304 stainless steel strip, rod, or tube
Electrolyte composition	$KOH_{(aq)}$
Electrolyte concentration	0.2 g/L
Electrolyte temperature	20–25 °C
Residence time	60–180 s
Voltage or electromotive force	6 V_{DC}

Source: Ref 1

necessary. If the mounting resin is not conductive, then clip the anode lead on the mount so that it directly touches the specimen. Be aware that a conductive resin could be etched preferentially to the specimen.

2. Immerse the anode lead and specimen in a 400 mL or larger beaker containing the electrolyte solution. Immerse the cathode in the beaker opposite the anode, such that the cathode does not touch the anode lead or specimen. Connect the two electrodes to the power source. The anode lead is typically a sharpened 3 mm diameter rod, and the cathode a similar rod with a 3 cm square plate welded on the end. Alternatively, a 6 mm diameter tube or 2 to 3 cm wide strip of stainless steel, copper, or platinum that is 2 cm longer than the depth of the beaker could be used as electrodes. The anode lead may be fashioned into a clip to maintain contact with the specimen. The anode lead and cathode should be of the same metal.
3. Set the parameters listed in Table 5.2 for the procedure being used. The parameters in Table 5.2 are also applicable to TiC, TaC, and B_4C, with some adjustment of the current or residence time. Some procedures are given in Table 5.1 and in Ref 1. Follow the instructions of the electrolytic kit if one is being used.
4. Actuate the power source and begin timing the process. When the specimen is adequately etched, clean and dry it as in Subroutine 3.1. Some electrolytically etched carbides are shown in Chapter 8.

Other Etching Methods

A number of other etching methods exist in addition to the thermal, chemical, and electrolytic methods. All etching methods are short on reproducibility and require a trial-and-error approach on the part of the ceramographer. No method is applicable to all ceramics, and no ceramic is easily etched by all methods. The etching methods and their applications are summarized in Table 5.1.

ASTM E 407, listed in Appendix A, contains some additional information regarding etching of microstructures.

Molten Salt Etching. Molten salts have been used to etch many ceramics as well or better than the thermal and chemical methods. Molten salts etch in a manner similar to the chemical etchants, except that the liquid solvent is not water (i.e., aqueous) but the electrolyte itself. Researchers have etched 99.995% alumina in molten vanadium pentoxide (V_2O_5) at 750 °C in less than 1 min (Ref 6). The V_2O_5 was contained in a nickel or platinum crucible, and the specimen was suspended in the molten salt on a platinum wire. The V_2O_5 residue was removed with hydrochloric acid.

Silicon carbide can be etched in molten $NaHCO_3$ (baking soda), $KHCO_3$, or $Na_2B_4O_7$ (dehydrated borax) in approximately 10 min. Alumina and sillimanite have been etched in molten KHF_2 in 10 min or less. The crucible was platinum, and the salt residue was removed with a phosphoric acid rinse (Ref 1). A disadvantage of molten salt etching is that the molten salts are highly reactive and their vapors very poisonous, in many cases.

A direct-sintered SiC plate etched in molten $NaHCO_3$ for 10 min at 950 °C is shown in Fig. 5.7. The ceramic and salt were heated together in a platinum crucible from room temperature in flowing argon, held at 950 °C for 10 min, and furnace cooled back to room temperature. The $NaHCO_3$ decomposed irreversibly to NaOH above 123 °C and melted at 846 °C. The solidified salt residue was removed with concentrated $HCl_{(aq)}$ in an ultrasonic bath for 3 min.

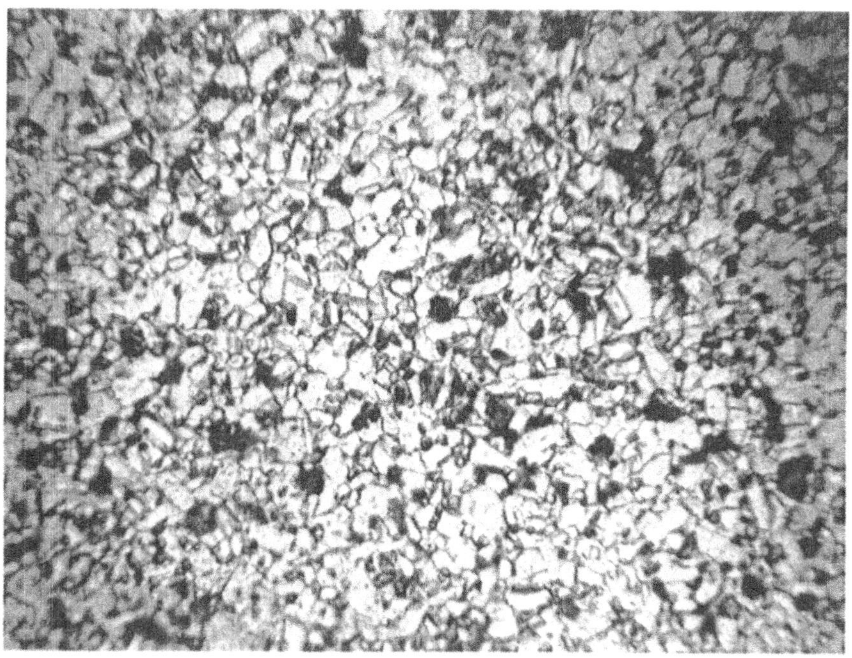

Fig. 5.7 Direct-sintered SiC etched in molten $NaHCO_3$ in a platinum crucible for 10 min at 950 °C in an argon atmosphere

Precious metal crucibles and wire for molten salt etching are available from Englehard Corporation in New Jersey. Refractory metal crucibles are available from Metal Technology, Inc. in Oregon.

Ion Etching. In ion etching, also known as *cathodic vacuum etching* or *ion bombardment,* argon or krypton ions in a 1 to 10 kV direct electric field at a vacuum pressure of 100 mtorr bombard and selectively attack grains of favorable crystallographic orientation by removing atoms. The specimen becomes the cathode in the reaction chamber. The rate of removal of atoms also depends on their atomic number and bonding state (Ref 1).

At the National Physical Laboratory in the United Kingdom, alumina and reaction-bonded SiC were ion etched with an accelerating voltage of 6 kV and a beam current of 30 μA for 45 min in an ion mill (Ref 2). A glassy sialon was etched somewhat less successfully at 7.5 kV and 50 μA for 5 h.

Partially stabilized ZrO_2 was ion etched with an accelerating voltage of 5 kV_{DC} and a beam current of 50 μA for 75 min in argon ions at an entry angle of 35° (Ref 11). The fine precipitates of monoclinic zirconia along the cubic grain boundaries were visible. The ion beam etching device shown schematically in Fig. 5.8 operates at an accelerating voltage of 1 to 7.5 kV_{DC}, a beam current up to 400 μA, an argon or krypton atmosphere at 2.5×10^{-4} mbar during entry, and an entry angle of 5 to 60°. The spec-

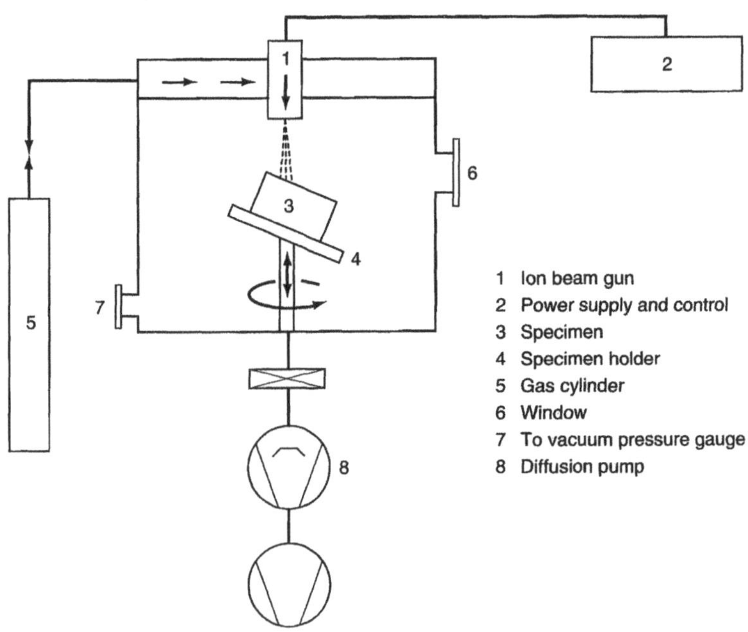

Fig. 5.8 Ion beam etching device. (Ref 11)

imen is not significantly heated by the ion beam in all cases and therefore does not necessarily require removal from its encapsulation.

In an early application of ion bombardment to ceramics, researchers ion etched urania, porcelain, and various cermets (Ref 10). All were etched with an accelerating voltage of 4 kV and a beam current of 2 mA in ionized krypton at 30 to 50 mtorr for 10 to 30 min. The nonconductive porcelain was modified with a 200-mesh stainless steel gauze overlay connected to the zirconium cathode plate. The Krypton ions were drawn through the apertures in the steel mesh onto the porcelain surface. A gray film that condensed on the porcelain where it was not covered by the gauze contained iron from the gauze and zirconium from the cathode plate.

Ion mills are available from Gatan Inc. in California, E.A. Fischione Instruments Inc. in Pennsylvania, and South Bay Technology Inc. in California.

Heat tinting is a means of surface discoloration as microstructural features of the specimen react at different rates while heated in air or other atmospheres (Ref 1). Titanium carbide can be heat tinted in approximately 5 min over a Bunsen flame. An important advantage of heat tinting is that it can be observed with the naked eye as it happens. The time and especially temperature of heat tinting in ceramography are much less than those of thermal etching. The contrast from heat tinting is likely due to interference layers created by oxidation-reduction phenomena rather than to the surface rounding associated with thermal etching. A heat-tinted TiC section is shown in Fig. 8.46.

Interference Layer. A vapor-deposited or sputter-coated interference layer of high refractive index on the polished surface of the specimen has a quasi-optical etching effect. Path differences in the reflected light between different microstructural features provide contrast through interference phenomena between those features, shown schematically in Fig. 5.9(a). Materials commonly used as coatings in the vacuum deposition method include TiO_2, SiO_2, ZrO_2, ZnS, SnO_2, Ta_2O_5, Nb_2O_5, and WO_4. Various metals are oxidized during deposition in the reactive sputtering method (Ref 18–20). Destructive and constructive interference caused by a refractive coating retard or enhance the reflections from phases that reflect approximately the same color and brightness without an interference layer.

An interference layer creates contrast in a microstructure by phase shifting the reflected rays from one ceramic compound relative to the reflection of an adjacent compound. The coating thickness, wavelength, absorption coefficients, and refractive indices are among the factors that affect the contrast. The layer thickness varies with the position on the specimen surface and the deposition equipment parameters. The incident beam from the microscope illuminator has a range of wavelengths, even if a color filter is used. Certain wavelengths are absorbed by each ceramic compound and by the interference layer. The intensities reflected by the coatings and ceramic phases may be quite different, depending on their respective refractive indices. The contrast comes from phases with different refractive indices and absorption coefficients reflecting different colors and brightnesses.

The interference layer in Fig. 5.9(b) is TiO_2 powder that was evaporated onto a reaction-bonded SiC heating element. The TiO_2 was compacted into a small pellet. A 3 mg TiO_2 pellet fragment was electrically heated in a spiral tungsten coil at 15 V_{AC} and 100 A, approximately 5 cm from the polished surface of the mounted SiC. The TiO_2 was deposited at a vacuum pressure of 50 mtorr for 30 to 60 s. Half the polished surface was masked with cellophane tape. The difference in contrasts at the boundary between the coated and masked halves is shown in Fig. 5.9(c).

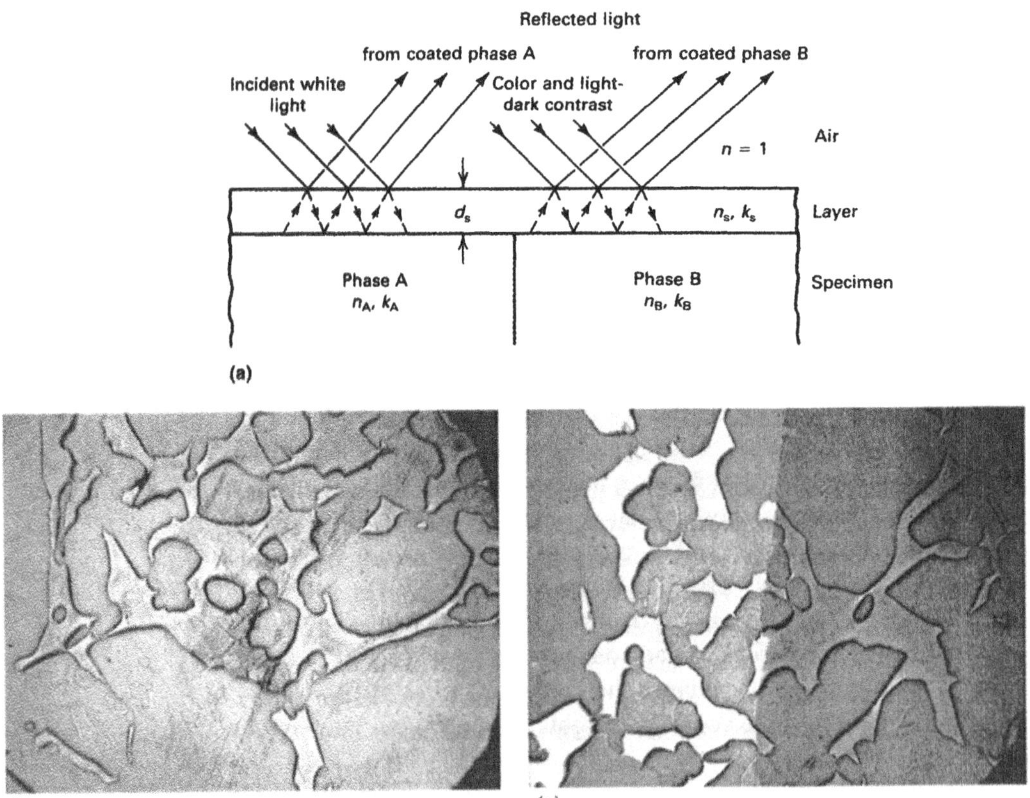

Fig. 5.9 Interference layers. (a) Contrast between phases A and B comes from optimization of the refractive index and absorption coefficient (n_s, k_s) of the layer with respect to the optical constants of the phases (n_A, k_A, n_B, k_B) and adjustment of the layer thickness d_s (Ref 17). (b) TiO_2 evaporative interference layer on a SiC heating element. (c) Interference layer (right) and as-polished (left) comparison on a SiC heating element. For color version of Fig. 5.9 (b) and (c), see endsheets of book.

In a special application, interference layers are used to prevent reflections from microscope lenses. A refractive coating of thickness $\lambda/4$ (or $3\lambda/4$, $5\lambda/4$, $7\lambda/4$, etc.) is applied to a microscope lens, where λ is the wavelength of the incident light. When the incident ray is reflected by the lens, it travels half a wavelength (twice $\lambda/4$) farther than and is shifted $\lambda/2$ out of phase with the light reflected by the interference layer. If the refractive index of the interference layer is equal to the square root of that of the lens glass ($\eta_{IL} = \sqrt{\eta_G}$; see Eq 8 in Chapter 7), then the intensities of the two reflected rays are the same, and the reflection is cancelled by destructive interference. More than one coating may be used to cancel the reflections of other wavelengths, because the incident beam is generally polychromatic.

A SiC varistor was etched by reactive sputtering with a platinum or iron cathode in an oxygen atmosphere (Ref 4). The ceramic contained large particles of α-SiC and SiO_2 in a porous matrix. Also, SiC bonded with free silicon and free carbon was etched by reactive sputtering with an iron cathode in oxygen. The SiC in the Si-SiC-C composite was first electrolytically etched in 10% oxalic acid at 15 V and the silicon chemically etched in 38 to 12 to 42 to 0.1 by volume HNO_3 + HF + CH_3COOH + iodine. Reactively sputtered interference layers, usually iron in oxygen, were used to enhance a number of composites, including alumina plasma sprayed on steel, plastic reinforced with glass fibers, aluminum reinforced with boron or SiC fibers, a joint between stainless steel and solder glass, a zirconia-perovskite-cermet fuel cell, and chromia plasma sprayed on nickel-clad steel.

The interference layer is very thin, less than 1 μm, and can be removed with relief polish. Evaporators and sputter coaters that deposit interference layers are available from Denton Vacuum LLC in New Jersey and Thermionics Vacuum Products in Washington.

Plasma etching requires specialized equipment not often found outside the semiconductor industry—a radio-frequency apparatus that rapidly attacks the silicon atoms in sialon, Si_3N_4, and SiC with a fluoride ion plasma from superheated CF_4 + O_2 gas. Plasma is an ionized, electrically conductive gas that is found in neon lights, fluorescent lights, lightning, and electric welding arcs. Plasma can be considered the fourth state of matter, after solid, liquid, and gas. In plasma etching, the CF_4 gas is admitted into a vacuum chamber and ionized in a high-frequency electric field. The fluorine free radicals in the resulting plasma react with silicon atoms in the ceramic to form SiF_4 gas. The SiF_4 gas is removed by the vacuum. The β-phase in sialon is selectively attacked by the plasma.

Plasma etching was used to reveal the microstructure of sialon metal-cutting tools (Ref 8). The sialon was dry pressed from a mixture of Si_3N_4, Al_2O_3, AlN, and Y_2O_3 and sintered at 1800 °C. The polished specimens

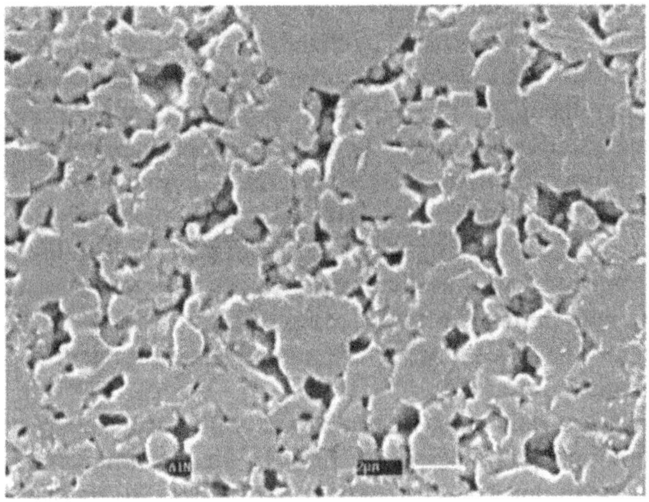

Fig. 5.10 Scanning electron microscopy photo of thermochemically etched, gold-coated AlN

were preheated for 10 min in the reaction chamber with nitrogen plasma at 62 Pa absolute pressure and 300 W forward power. The power supply was operated at an industry-standard frequency of 13.56 MHz. The nitrogen was evacuated and replaced with CF_4 + 5% O_2 at 62 Pa and 400 W for 30 to 45 s. A yttria-rich vitreous phase sharply contrasted with the β-sialon phase in the microstructure after the plasma etch. Additional etching time only roughened the surface and deteriorated the appearance of the microstructure.

Graphite was etched in O_2 flowing at 50 cm³/min per cm² of surface (Ref 4). The pressure in the chamber was 70 Pa and the power 200 W for 20 to 60 min. It was reported that plasma etching is effective on SiC only when the SiC has at least 5% unspecified additives (Ref 21).

Plasma etching equipment is available from Oxford Instruments PLC in England, PETS Inc. in California, Plasma Etch Inc. in Nevada, Process Integration LLC in Arizona, and Trion Technology in Arizona.

Thermochemical Etching. A combination of chemical and thermal etching methods was used to thermochemically etch 96% alumina, aluminum nitride, and nickel-zinc-ferrite ($Ni_{0.5}Zn_{0.5}Fe_2O_4$) ceramics (Ref 7). The ceramics were first chemically etched 2 to 3 min in mixtures of 0.8 N orthophosphoric acid and 0.5 N nitric acid, then thermally etched for 10 min at approximately 70% or less of their respective absolute sintering temperatures. The Al_2O_3 was sintered at 1630 °C, polished, soaked in a 3 to 1 volume ratio of the two acids, and etched at 1150 °C. The AlN was sintered at 1900 °C, polished, soaked in a 2 to 1 volume ratio of the two acids, and etched at 1100 °C. The ferrite was sintered at 1240 °C, polished, soaked in a 1 to 1 volume ratio of the two acids, and etched at

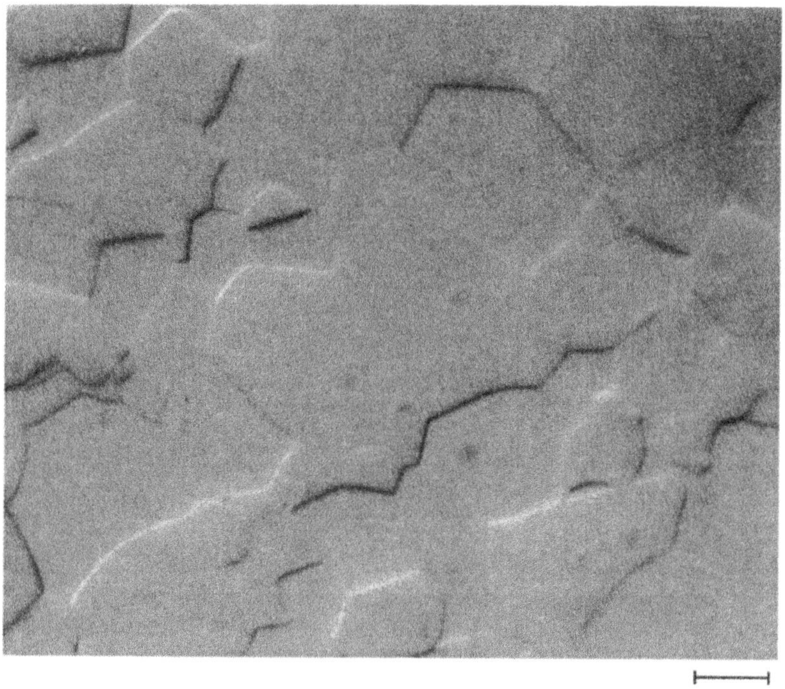

Fig. 5.11 Hot isostatically pressed $MgAl_2O_4$ spinel relief polished with 0.05 μm γ-Al_2O_3 and viewed in differential interference contrast. The mean grain size is 72 μm by the circle-intercept method.

850 °C. In each case, the thermochemical delineation was superior to thermal etching alone.

Thermochemical etching is purported to cause less surface modification and grain growth, due to the lower severity of the chemical etch compared to the boiling etchant methods and the lower thermal etching temperatures. An AlN modulus-of-rupture bar that was vibratory polished 8 h with colloidal silica and thermochemically etched is shown in Fig. 5.10. The bar was thermally etched in an atmosphere of flowing N_2.

Relief Polish. Relief polishing, which could also be considered mechanical etching, is mentioned briefly in Chapter 4 as an optional final step in grinding and polishing. Relief between adjacent grains, caused by differential abrasion rates between harder and softer phases, enhances the contrast in the microstructure. The advantage of relief polish is its simplicity. The disadvantages include: the contrast is not nearly as good as what can be achieved by most other etching methods, relief causes focusing difficulties at higher magnifications, and relief polishing tends to round the specimen edges. Relief polishing abrades the intergranular glass in 85 to 98% alumina compositions but has little effect on the alumina grains. Relief polishing abrades the softer silicon matrix in reaction-bonded SiC but has little effect

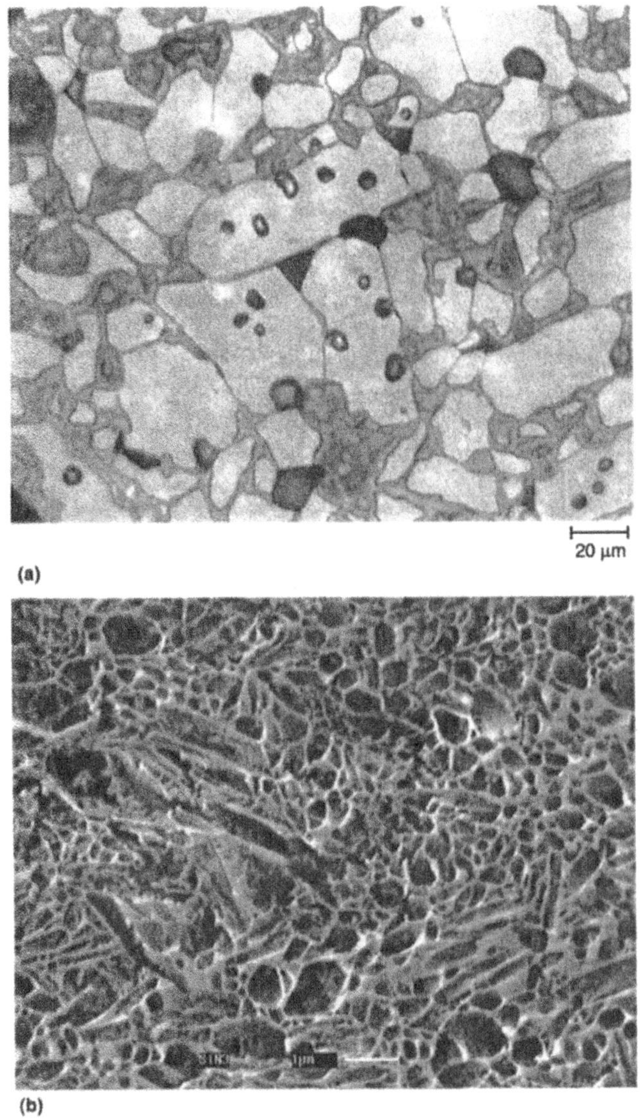

Fig. 5.12 Overetching. (a) Overetched 98% Al_2O_3. (b) Scanning electron microscopy photo of overetched Si_3N_4. The Si_3N_4 was plasma etched 1 h at 50 mtorr in flowing O_2 at 10 cm^3/min and CF_4 at 60 cm^3/min. The inductively coupled plasma head was set at 800 W, the sample chuck at 150 W. Courtesy of D.A. Chinn, Sandia National Laboratory

on the harder SiC grains. The effect of relief polishing is quite dramatic in $MgAl_2O_4$ spinel, as shown in Fig. 5.11, and can be used prior to thermal or chemical etching for optimal grain-boundary delineation. Very brief relief polishing is also useful for polishing the metal components in cross sections of microelectronic devices and other metallized ceramics.

To relief polish a ceramographic specimen, use 0.3 μm (sometimes sold as Linde A) or 0.05 μm (Linde B) γ-Al_2O_3 particles suspended in distilled water on napped (fuzzy texture) polishing cloth on a manual or automatic polishing wheel or vibratory polisher. Colloidal silica can also be used and may have an attack-polishing effect on some ceramics, as it does on many metals. Relief polishing on a polishing wheel, manual or automatic, usually takes 5 min or less. Relief polishing on a vibratory polisher can take several hours.

Specimens for microhardness testing or porosity measurement should not be relief polished. Very fine-grained ceramics that must be viewed under high magnification in visible light should not be relief polished.

Overetched Ceramics

Ceramographic specimens that have been overetched must be remounted and reground, in most cases. Electrolytic etching can usually be removed with relief polish. Overetching can cause a number of artifacts in the microstructure, including:

- Grain growth of the primary phase in thermally etched ceramics
- Dissolution or precipitation of secondary phases in thermally etched ceramics, as shown in Fig. 5.12(a)
- Degradation of the primary phase or other features of interest in chemically etched ceramics
- Etch pits and enhancement of scratches and pluckouts in chemically etched ceramics
- Rough surface in plasma-etched silicon-bearing ceramics, as shown in Fig. 5.12(b)
- Edge rounding in relief-polished ceramics
- Discoloration and loss of contrast in electrolytically etched ceramics

REFERENCES

1. G. Petzow, *Metallographic Etching,* 2nd ed., ASM International, 1999, p 38–50 and 153–174
2. D.J. Clinton, *A Guide to Polishing and Etching of Technical and Engineering Ceramics,* The Institute of Ceramics, 1987, p 17–21 and 27–34
3. L.A. Lay, *Corrosion Resistance of Technical Ceramics,* Her Majesty's Stationery Office, 1983, p 31–32
4. G. Elssner, H. Hoven, G. Kiessler, and P. Wellner, *Ceramics and Ceramic Composites: Materialographic Preparation,* R. Wert, Trans., Elsevier Science Inc., 1999, p 40–48, 89, 95

5. H. Mörtel, Microstructural Analysis, *Ceramics and Glasses,* Vol 4, *Engineered Materials Handbook,* ASM International, 1991, p 570–579
6. J.D. Katz and G. Hurley, Etching Alumina with Molten Vanadium Pentoxide, *J. Am. Ceram. Soc.,* Vol 73 (No. 7), July 1990, p 2151–2152
7. I.O. Owate and R. Freer, Thermochemical Etching Method for Ceramics, *J. Am. Ceram. Soc.,* Vol 75 (No. 5), May 1992, p 1266–1268
8. C. Chatfield and H. Norström, Plasma Etching of Sialon, *J. Am. Ceram. Soc.,* Vol 64 (No. 9), Sept 1983, p C-168
9. N.P. Bansal and C.H. Drummond III, Kinetics of Hexacelsian-to-Celsian Phase Transformations in $Sr_2Al_2Si_2O_8$, *J. Am. Ceram. Soc.,* Vol 76 (No. 5), 1993, p 1321–1324
10. T.K. Bierlein, H.W. Newkirk, Jr., and B. Mastel, Etching of Refractories and Cermets by Ion Bombardment, *J. Am. Ceram. Soc.,* Vol 41 (No. 11), June 1958, p 196–200
11. G. Petzow, R. Paris, E.D. Albrecht, and J.L. McCall, Ed., *Microstructural Science,* Vol 9, Elsevier North Holland, Inc., 1981, p 83–90
12. R.E. Chinn, Preparation of Microstructures of Alumina Ceramics, *Structure,* Vol 33, 1998, p 16–20
13. R.Z. Chen and W.H. Tuan, Thermal Etching of Alumina, *Am. Ceram. Soc. Bull.,* Vol 79 (No. 10), Oct 2000, p 83–86
14. E.M. Levin, C.R. Robbins, and H.F. McMurdie, *Phase Diagrams for Ceramists,* 2nd ed., The American Ceramic Society, 1969, p 140
15. E.M. Levin, C.R. Robbins, and H.F. McMurdie, *Phase Diagrams for Ceramists, 1969 Supplement,* The American Ceramic Society, 1969, p 88
16. R.E. Chinn, Grain Sizes of Ceramics by Automatic Image Analysis, *J. Am. Ceram. Soc.,* Vol 77 (No. 2), 1994, p 589–592
17. G. Petzow and G. Elssner, Etching, *Metallography and Microstructures,* Vol 9, *Metals Handbook* (9th ed.), American Society for Metals, 1985, p 60
18. G.F. Vander Voort, *Metallography: Principles and Practice,* American Society for Metals, 1984, p 187–190
19. R.S. Crouse, Interference Films by Vacuum Deposition, *Metallography and Microstructures,* Vol 9, *Metals Handbook* (9th ed), American Society for Metals, 1985, p 147–148
20. H. Hoven, K. Koizlik, J. Linke, and H. Nickel, Interference Films by Reactive Sputtering, *Metallography and Microstructures,* Vol 9, *Metals Handbook* (9th ed.), C.G. Goetzel, Trans., American Society for Metals, 1985, p 148–150
21. U. Täffner, Ceramography—an Exciting Area in Materialography, *Structure,* Vol 33, 1998, p 3–6

CHAPTER 6

Petrographic Thin-Section Preparation

THIN SECTIONS, also known as petrographic sections because they are widely used by geologists to examine minerals, are ceramographic sections that are sawed and ground thin enough to transmit visible light. Etchants are not used—the contrast comes from the refraction of the transmitted light by optically anisotropic, noncubic, translucent grains in polarized light. Translucent cubic ceramics and glass are optically isotropic but can also be viewed in transmitted light. Cubic ceramics have little, if any, contrast between grains and appear dark in transmitted light through crossed polars. Petrography is applicable to α-Al_2O_3, which has a hexagonal crystal structure.

The advantage of thin sections over ceramographic "thick" sections is fast, easy preparation and evaluation, which precludes polishing and etching. The disadvantages of thin sections include the difficulty of automated preparation and evaluation and the semiquantitative nature of the evaluation. Hand-grinding of thin sections requires a certain amount of finesse and dexterity that can only come from practice.

A procedure for preparation of thin sections of concrete, ASTM C 856, is listed in Appendix A. Thin sections are well suited for viewing all grades of alumina. Some examples of alumina thin sections are shown in Chapter 8. This Chapter describes only the preparation of thin sections. Transmitted light and polarization are discussed in Chapter 7.

Sawing

Cut a section 1 mm thick and approximately 10 mm in diameter, as described in the section "Sawing" in Chapter 3 and as shown in Fig. 6.1. A rectangular or irregular cross section that is no more than approximately 10 mm on any side is also suitable. An automatic saw that makes flat, parallel cuts

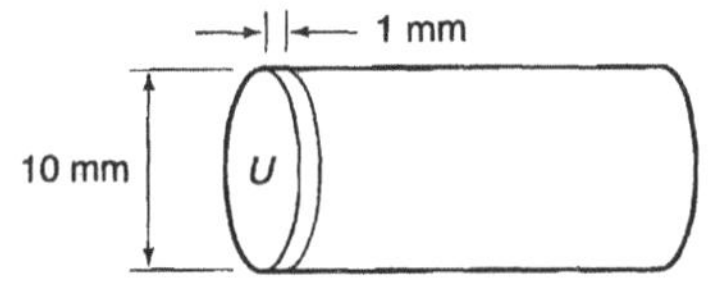

Fig. 6.1 Saw a section approximately 1 mm thick and 10 mm in diameter. The sawed surface must be flat. The letter "U" indicates the up side of the specimen and should be designated with a pencil mark.

with a diamond wafering blade yields the best results. A flat, smooth face on the specimen is essential, because this face is glued to a microscope slide. Some equipment manufacturers make a saw especially for thin-section preparation.

Mounting

One face of the specimen must be ground smooth before the face is glued to a microscope slide. The slide facilitates preparation and viewing of the specimen. Specifications for cover glasses and slides are given in ASTM E 211, which is listed in Appendix A.

A fragile or friable specimen can be embedded in castable resin, as described in the section "Mounting" in Chapter 3, before it is glued to the slide. Fill the mold shown in Fig. 3.6 only to the top of the specimen. The castable resin protects the specimen from disintegrating during the grinding steps described in the section "Grinding" in this Chapter.

Thin-Section Mounting Method. The steps are:

1. Mark the rougher face of the specimen with a pencil. Grind the smoother face on a 200-grit or finer metal-bonded diamond grinding wheel lubricated with a continuous stream of water. The eraser end of a pencil is useful for holding the small section against the abrasive

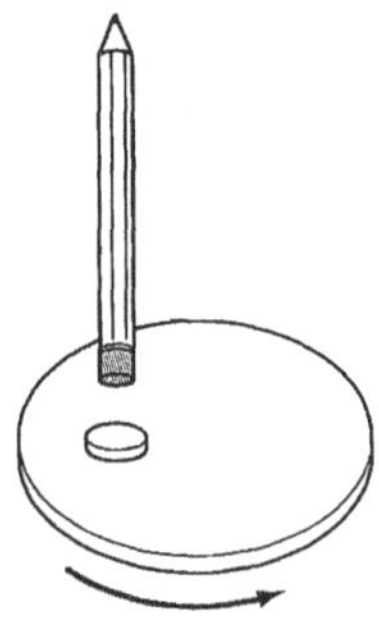

Fig. 6.2 Smooth the sawed face on successive abrasive wheels. The "U" face is up, away from the abrasive. A pencil eraser is useful for holding the sawed face against the rotating abrasive wheel.

wheel, as in Fig. 6.2. Grind the smoother face on successively finer wheels, 400- and 600-grit or equivalent, and be sure to thoroughly clean the section after each grinding step. After the final grinding step, clean and dry the specimen, as in Subroutine 3.1.

2. Prepare a petrographic (46 by 26 mm) microscope slide. If the slide has a frosted edge, write the specimen identification in the frosted area with a fine felt-tip pen or a 0.5 mm pencil. If the slide does not have a frosted edge, write on the smooth glass with a felt-tip pen or frost an edge by grinding it for a few seconds on a 200-grit wheel. Alternatively, an engraving pen can be used to label the slide. Be sure the slide is clean and dry before proceeding. Place the slide, labeled side down (so that the label is away from the glued specimen), on a hotplate heated to approximately 80 °C.
3. Mix the liquid epoxy resin and hardener on another glass slide with a clean laboratory spatula, allowing at least 1 mL of epoxy for each specimen. Several slides can be prepared at one time. Epoxy resins with the correct refractive index are available from Hillquist or Logitech.
4. Smear a dab of the mixed epoxy onto the smooth, ground face of the specimen and press the specimen firmly onto the center of the petrographic slide, as in Fig. 6.3. Only a very thin layer of epoxy, without any bubbles, should separate the specimen from the slide. The specimen should not be floating in a pool of epoxy.
5. Cure the epoxy according to the manufacturer's instructions, typically by heating the slide on the hotplate at 80 °C for at least 30 min.

Grinding

The epoxy must be completely cured and without any bubbles between the specimen and the slide, prior to the grinding operations. In lieu of

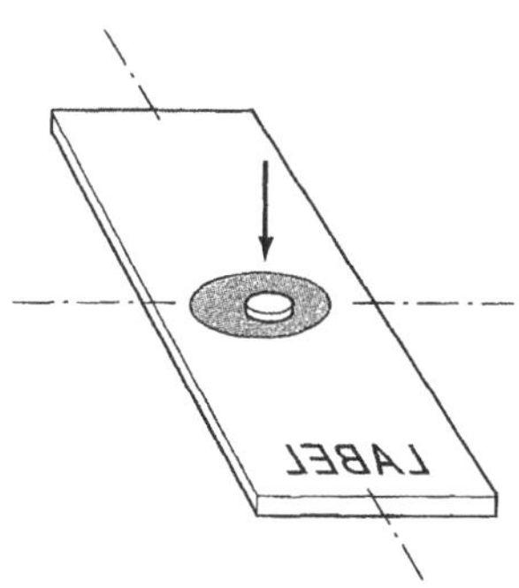

Fig. 6.3 Label the slide on one side, clean the slide as needed, and place the labeled side down (which is why the handwritten "LABEL" is backwards) on a hotplate. Daub the mixed epoxy on the smooth face opposite the "U" side of the specimen and press the specimen firmly onto the center of the slide, with the "U" side up. The excess epoxy on the specimen perimeter is ground off in a subsequent step.

grinding or lapping, the specimen can also be sawed, as described in step 4 that follows. The grinding or sawing reduces the thickness of the specimen until it can transmit visible light.

Thin-Section Grinding Method. The steps are:

1. Clamp the slide in a petrographic slide holder, shown in Fig. 6.4(a) and (b). Coarse-grind the specimen on a 200-grit, metal- or resin-bonded diamond disc until the specimen becomes translucent and resembles waxed paper. Use flowing water as the grinding lubricant. Thoroughly rinse off all the grinding residue under tap or distilled water. Some of the equipment manufacturers listed in Appendix B offer slide holders. If the excess epoxy interferes with the grinding, the excess can be removed by scoring around the specimen with a razor and shaving off the excess.
2. Fine-grind or lap the specimen on 400- and 600-grit diamond discs until the specimen is approximately 30 μm thick or until fingerprints can be seen through the back of the slide as the specimen is pressed. At this point, the slide is generally adequate for inspection at 200× or lower magnification. Further fine grinding or polishing is optional, depending on how much detail is to be seen in the thin section. Rinse and dry the entire slide.
3. The thickness can be determined semiquantitatively from the Michel-Lévy chart in Fig. 6.5 and Eq 1 (Ref 1). The desired thickness is approximately 30 μm. Observe the slide in transmitted light with crossed polars but without a color filter and note the colors of the grains. Equation 1 relates the retardation color or optical path difference to the thickness of the crystal and its birefringence:

$$t = \frac{r}{|\eta_\varepsilon - \eta_\omega|} \qquad \text{(Eq 1)}$$

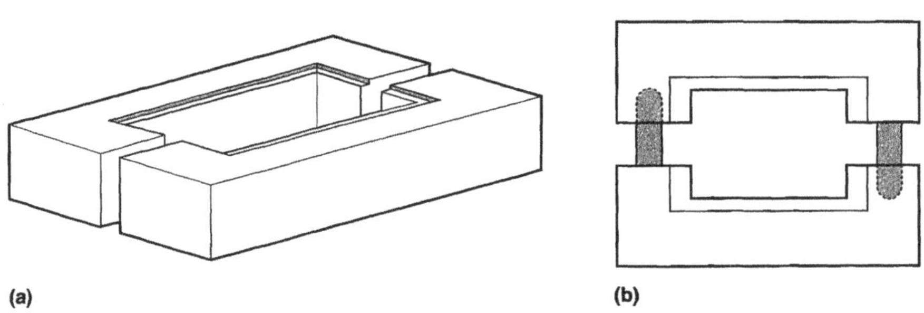

Fig. 6.4 Thin-section slide holder. (a) The thin-section slide holder consists of two interlocking plastic pieces that clamp the slide in the recess of one face. The slide is slightly thicker than the depth of the recess. (b) Two copper pins align the interlocking plastic pieces of the thin-section slide holder.

where t is the thickness of a birefringent crystal (nm); r is the retardation color, expressed as the optical path difference of a double-refracted beam of light (nm); $|\eta_\varepsilon - \eta_\omega|$ is the birefringence of the crystal; and η_ε and η_ω are the refractive indexes of the crystal (see the section "Crystal Optics" in Chapter 7). The birefringence of α-Al_2O_3 is approximately 0.008. For a 30 μm thick alumina section, the retardation would be 240 nm. The interference color corresponding to 240 nm is gray-white, from the Michel-Lévy chart. If the grains are bright yellow or brown-yellow, the section may still be too thick at 40 to 50 μm.

4. An alternative to grinding is sawing, and a few saws are intended specifically for thin sections. Clamp the slide in a vacuum chuck (Fig. 3.2a). The gasket of the vacuum chuck may need to be wetted in order

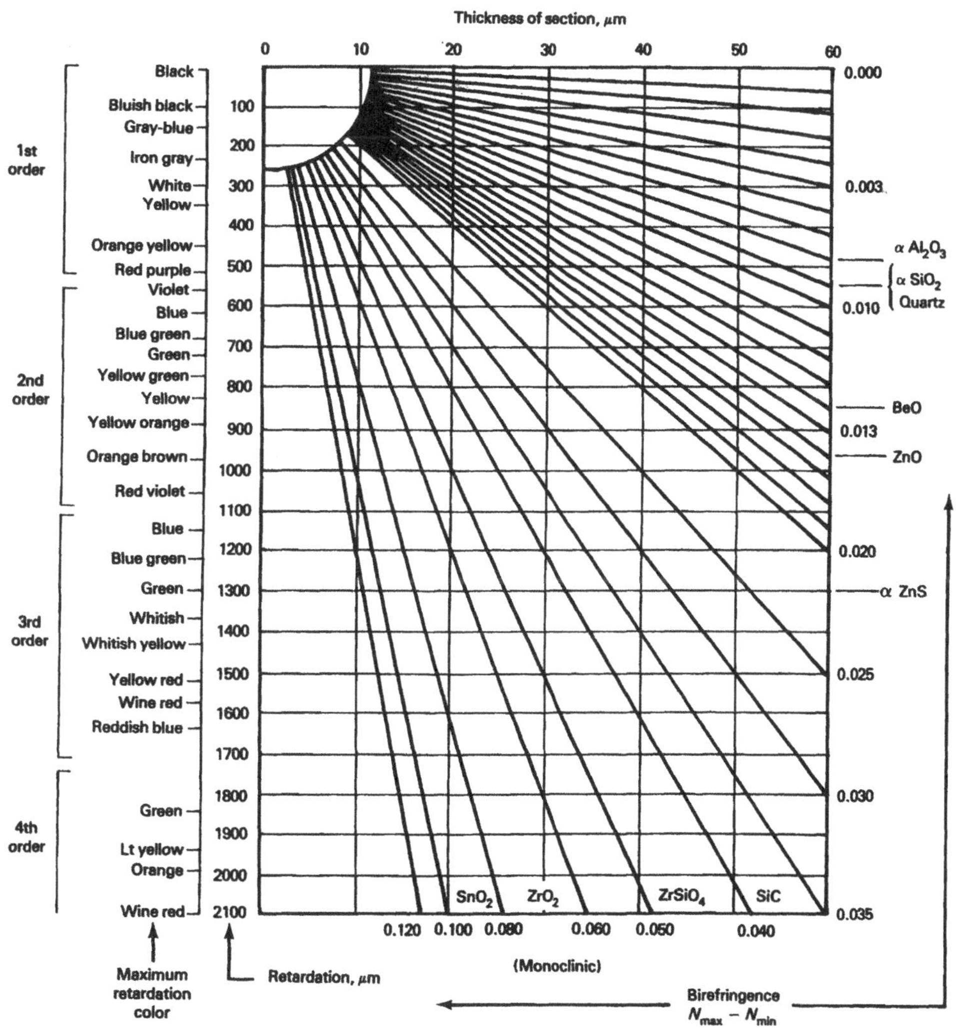

Fig. 6.5 Michel-Lévy chart. The abscissa represents retardation, and the color sequences are due to Newtonian interference phenomena. For a different color version of Fig. 6.5, see end sheets of book.

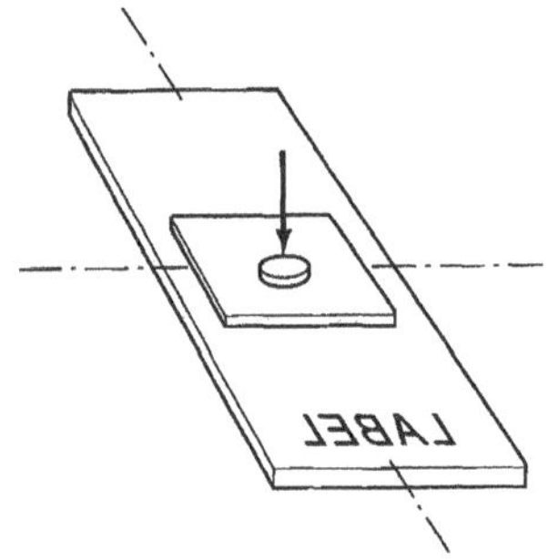

Fig. 6.6 Grind through (or saw off) the "U" until the specimen is approximately 30 μm thick. Clean and dry the slide and attach a cover glass on the specimen with a drop or two of Canada balsam dissolved in xylene. Gently press the cover glass onto the specimen with a clean tool.

to maintain the vacuum. Saw the specimen to a thickness of approximately 0.030 mm under low pressure and medium speed. Rinse the entire slide to remove the swarf. Fine-grind the slide as necessary, as in step 2, to achieve the final thickness.

5. An alternative to grinding or sawing is ion milling, used primarily to prepare specimens for transmission electron microscopy (TEM). Ion mills are mentioned in the section "Other Etching Methods" in Chapter 5 (under "Ion Etching"), and in Chapter 7 in conjunction with TEM under the heading "Other Microscope Types."
6. Place the slide on a hotplate at 80 °C. Daub the specimen with Canada balsam dissolved in xylene (C_8H_{10}, xylol, or dimethylbenzene) when the slide is dry. Set a cover slip of thickness that agrees with the microscope objective—typically No. 1 (0.13 to 0.16 mm) thickness, 18 by 18 mm—on the pool of Canada balsam and press lightly with a clean tool to remove any bubbles, as in Fig. 6.6. Heat the slide for 5 min or until the solvent evaporates and the cover slip is held firmly. Lakeside 70 cement (Lakeside Microscope Accessories Inc.), or the epoxy used for mounting, can be used instead of the Canada balsam. Canada balsam is an oleoresin (turpentine) secreted by balsam fir trees and is available from most microscope accessory vendors. Canada balsam has a refractive index of 1.54, approximately the same as glass.

At this point, the slide is ready for viewing in transmitted light, which is discussed in the section "Petrography—Transmitted Light and Thin Sections" in Chapter 7. If necessary, the cover slip can be removed by redissolving the Canada balsam with a few drops of xylene.

REFERENCE

1. W.C. McCrone, L.B. McCrone, and J.G. Delly, *Polarized Light Microscopy,* McCrone Research Institute, 1984, p 147–149

CHAPTER 7

Optics and Microscopy

VARIOUS TECHNIQUES of ceramographic microscopy, which some authorities call optical etching, are discussed in this chapter. To the beginning microscopist, the microscope has only two important controls: the illuminator switch and the focus knob. A professional-quality microscope affords the advanced microscopist many more controls and accessories for optimal imaging and quantitative measurements. Much of the microscope jargon is defined in ASTM E 7 and E 175, listed in Appendix A. A semimonthly journal of microscopy that covers a wide variety of instruments, techniques, and materials is *Microscopy and Analysis,* published by Rolston Gordon Communications in England.

Some other types of imaging equipment are discussed in this chapter, including electron microscopes and acoustic microscopes, and some alternative viewing and preparation techniques are described.

The Microscope

The two basic types of microscope used in ceramography are upright and inverted. The inverted configuration is common to metallographs, where the mounted specimen is face down on the microscope stage and both the light source and the objective are below the stage. The specimen is always level but cannot be viewed in transmitted light. The upright configuration is required for transmitted-light microscopy, but in the reflected-light mode, the specimen must be leveled, as described in Subroutine 7.1. The nomenclature of an upright microscope is shown in Fig. 7.1.

The *objective* in Fig. 7.1 is a cylindrical device that contains an upper and lower lens and provides most of the magnification of the microscope. Focus depends on the distance between the objective and the specimen, the working distance. The objectives slide into position or are threaded into a turret to allow for quick changes of magnification range. The nom-

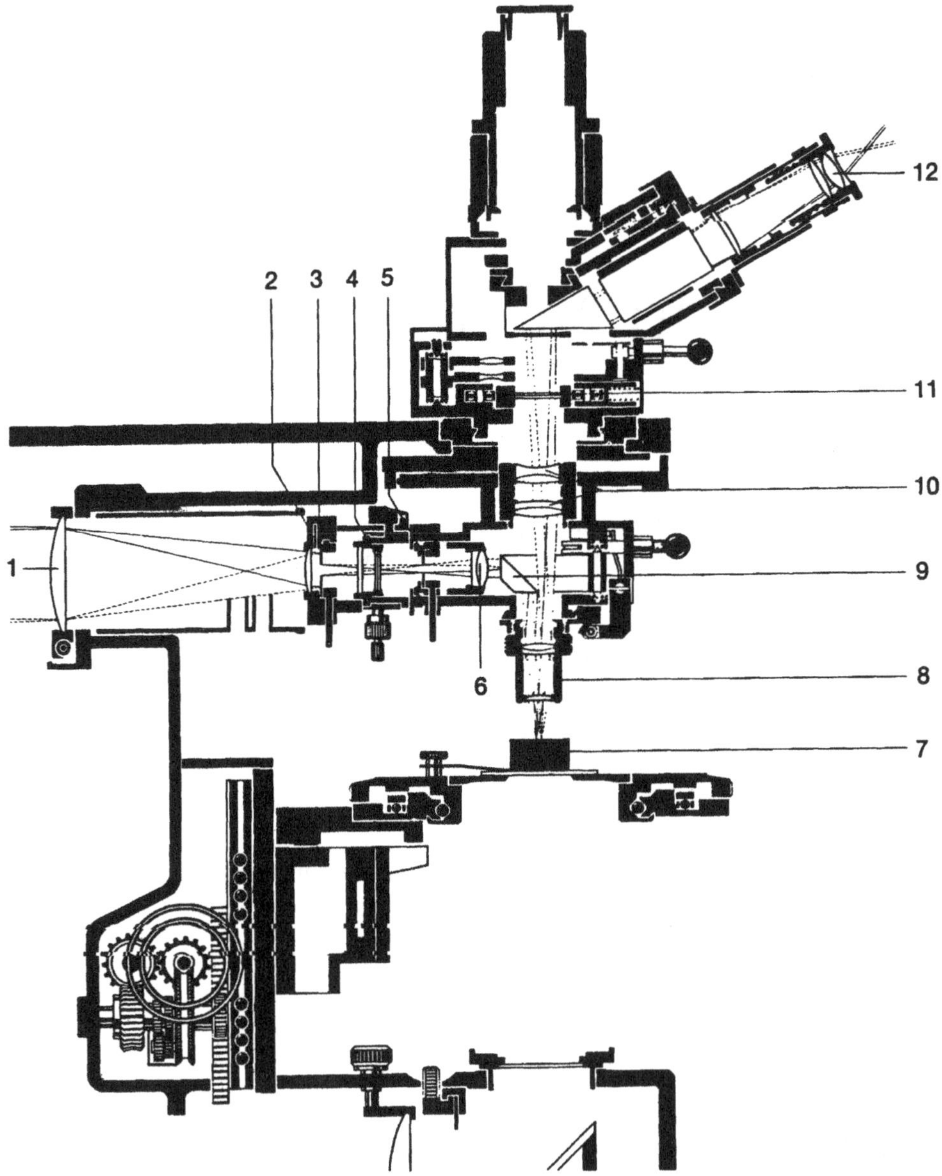

Fig. 7.1 Upright Leitz microscope showing the light path and key components. 1, hinged lens; 2, half stop; 3, aperture diaphragm; 4, filter or prism polarizer; 5, field diaphragm; 6, centrable lens, used to center the field diaphragm; 7, polished section; 8, objective; 9, compensating prism, with switchover against optical-flat reflector; 10, tube lens (intermediate optical system); 11, rotating analyzer; 12, eyepiece with focusing eyelens. Source: Ref 1

inal magnification, working distance in millimeters, and numerical aperture or angular aperture are usually printed on the side of the objective. Some objectives contain a Nomarski-modified Wollaston prism for differential interference contrast, which is described in a later section in this chapter. Some objectives are intended to be used only with immersion oil (see the section "Oil Immersion" in this chapter) and may have "oil" (or *oel* on some German brands) printed on the side. Some objectives can only be used in dark-field (see the section "Dark-Field Illumination" in this Chapter). Spare objectives should be stored in a dust-tight, padded objective box when not in use. The objective is held in place by the nosepiece, which can be a revolving turret that holds several objectives.

The *polarizer* and *analyzer* are both prisms that transmit only light waves of a certain orientation. This type of directional filtering is called polarization and is used in some sunglasses and traffic signal lenses. In polarized-light microscopy, light from the illumination source is first polarized by the polarizer, projected onto or through the specimen, and polarized a second time by the analyzer before it enters the oculars. The analyzer is contained in the body tube. The analyzer polarization direction is 90° from the polarizer (i.e., crossed polars) in polarized-light microscopy, an optical etching technique. Contrast between optically anisotropic grains in polarized light is enhanced in the form of color and intensity differences due to path-length differences caused by crystallographic orientation or relief.

The diaphragms are adjustable apertures that restrict the amount of light by changing the diameter of the light path. The *field diaphragm* limits the size of the field of view and is usually opened enough so that the aperture edges are just outside the area seen by the cameras. The *iris diaphragm* in transmitted-light mode controls the size of the light beam into the *substage condenser.* The substage condenser uses lenses to concentrate the light beam onto a thin-section slide. The diaphragms are opened and closed by levers. A diaphragm that is open too wide reduces image quality with excess brightness and interference halos. A diaphragm that is not open wide enough can make the image too dark or the field too small.

The *oculars* are the eyepieces where one looks into the microscope. Oculars typically magnify the image in the objective by a factor of ten. A Huygenian ocular has an upper and lower plano-convex lens and forms a real image between the lenses. A plano-convex lens is flat on one side and curved outward on the opposite side. The two oculars on a binocular microscope must be focused independently on the objective image for the left and right eyes. If two or more people use the same microscope, the ocular focal settings will differ for each user. A *reticle*—a transparent glass disc engraved with a measurement scale—may be inserted between the lenses of one of the oculars. Some reticles have parallel line segments similar to the stage micrometer in Fig. 7.3, and some have a grid similar to the overlay in Fig. 9.5(c). Others have various-size circles for measurement of asbestos fibers.

All lenses have some degree of aberration, that is, defects that reduce resolving power and distort images. A light beam through a lens with *spherical aberration* does not converge at a sharp focal point but in a finite space. The image of a narrow beam consists of fuzzy, concentric halos caused by diffraction. *Achromatic* objectives are corrected for spherical aberration but only in a narrow band of the color spectrum, such as yellow-green. A yellow-green filter (wavelength $\lambda \approx 540$ to 550 nm) is often inserted into the light path to sharpen the image. A light beam through a lens with *chromatic aberration* is refracted such that different colors converge at different locations. *Apochromatic* objectives are corrected for chromatic and spherical aberrations over a spectral range with lenses of fluorite (CaF_2) and special glass compositions. *Astigmatism* is length-width distortion of objects, mostly on the perimeter of the image, caused by incorrect curvature in the lens.

A number of illumination sources are available for microscopes, each with various advantages, disadvantages, and special applications compared to the others. Some can be operated only at maximum intensity, while others have variable intensity adjusted by a rheostat. Mercury vapor lamp, carbon arc lamp, tungsten-halogen lamp, xenon lamp, and fiber optic illuminator are among the choices. With a special filter, the mercury vapor lamp can provide ultraviolet illumination for fluorescence and increased resolution. The carbon arc lamp provides high-intensity white light in alternating or direct current. The tungsten lamp is highly regarded for black-and-white photography. A fiber optic accessory can provide oblique (not parallel to the direction of view) illumination. At least one spare bulb should always be kept on hand near the microscope. The bulb should be handled with a tissue or gloved hand to prevent fingerprints on the glass.

Subroutine 7.1: Leveling a Specimen for Reflected-Light Viewing in an Upright Microscope

This procedure is unnecessary for an inverted microscope or an upright microscope with a self-leveling stage:

1. Place a ball of modeling clay, such as Plasticine, approximately 10 mm in diameter, on a microscope slide. Modeling clay is inexpensive and available from most hobby stores.
2. Use a leveling press, available from most manufacturers of upright microscopes, to press the specimen onto the clay. Place a tissue on top of the specimen surface if the upper press surface is oily or dirty. The clay should be sticky enough to adhere the specimen to the slide. Modeling clay can be reused many times until it gets contaminated or dry and brittle.
3. Set the slide with the mounted specimen on the microscope stage and secure it with the stage clamps.

Microscopy is an old subject, and many reference texts have been written. Some are of a general nature, while others address a more specific field, such as metallography, mineralogy, or biology. Books by Vander Voort, Rochow and Tucker, and McCrone et al. address microscopy in the context of microstructures or materials science.

Köhler Illumination

Ceramographic image quality and usefulness depend on illumination as well as quality of preparation. Most professional-quality microscopes can be adjusted for Köhler (pronounced "curler") illumination, as shown schematically for reflected light in Fig. 7.2. Köhler illumination is applicable to both reflected and transmitted light, optimizes resolution and contrast, uniformly distributes incident light intensity on or through the specimen, and uses any type of lamp filament. Köhler illumination, devised by August Köhler in the 1890s, requires the microscope to have a lamp-condensing lens and a lamp iris.

To apply Köhler illumination to a transmitted-light microscope, focus and center the light source (filament) image in the substage iris diaphragm. Focus and center the field diaphragm image in the object plane. Focus and center the substage iris image in the back focal plane of the objective, as shown in Fig. 7.2(a).

To apply Köhler illumination to a reflected-light microscope, as in Fig. 7.2(b), bring the specimen into focus. Close the illumination condenser aperture or lamp field stop to its minimum diameter; its edges should be in focus when the specimen is in focus. Center the illumination condenser aperture in the field of view, usually with levers or thumbscrews. Open the illumination condenser aperture to the edges of the field of view but no farther. Switch the Bertrand lens into the optical path and focus the Bertrand lens on the radiant field stop or aperture stop. Narrow and center the radiant field stop, then open it until it just fills the field of view. If the microscope does not have a Bertrand lens, remove one of the oculars and observe the radiant field stop in the objective pupil.

To center the filament in transmitted or reflected light, reduce the intensity of the light source to approximately 50% of full power, swing the collector lens or lamp condenser out of the light path, switch the Bertrand lens into position, and adjust the illuminator position to sharpen and center the filament image in the back focal plane of the objective. The illuminator position is usually controlled by thumbscrews on the illuminator housing. Return the collector lens to the light path, switch the Bertrand lens out of the light path, and return the intensity to optimal contrast and eye comfort.

If the microscope does not have a Bertrand lens, reduce the intensity of the light source as necessary, swing the collector lens out of the light path,

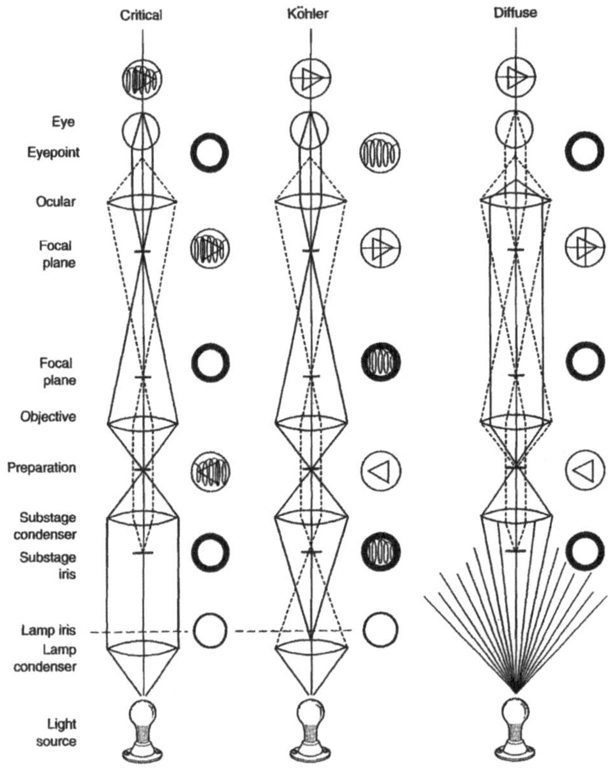

(a)

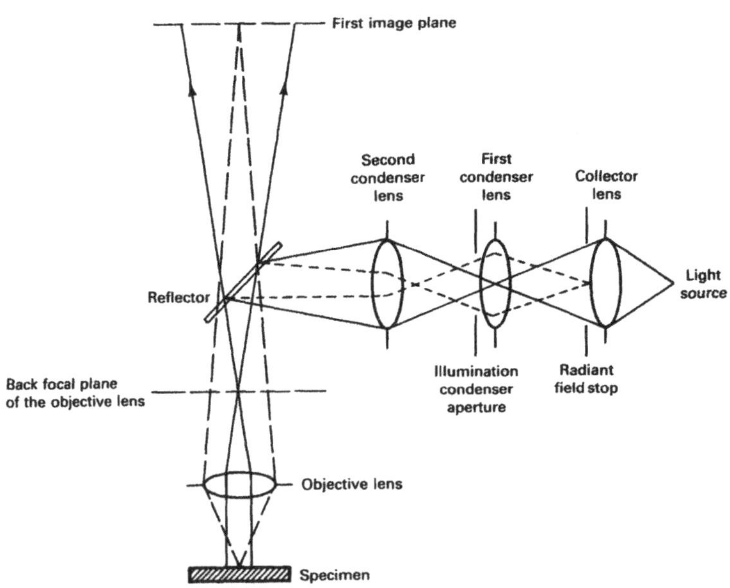

(b)

Fig. 7.2 Köhler illumination. (a) Nelsonian (critical), Köhler, and diffuse illumination in a transmitted-light microscope (Ref 2). The specimen is labeled as "Preparation." (b) Köhler illumination in a reflected-light microscope. Source: Ref 3

remove the condenser lens assembly or substage condenser from the microscope, and observe the filament through the radiant field stop or lamp iris. A tissue over the radiant field stop or lamp iris forms a projection of the filament. Adjust the illuminator position to sharpen and center the filament image on the tissue. Remove the tissue and replace the condenser lens assembly or substage condenser and the collector lens. Increase the intensity as necessary for maximum contrast and eye comfort. The filament position in the tissue projection may not agree precisely with the Bertrand lens but still yields a uniformly illuminated image.

In some Zeiss microscopes, the Bertrand lens in the body tube is designated with "PH" on the upper Optovar thumbwheel. The lower Optovar thumbwheel is used to focus the Bertrand lens. Microscope nomenclature varies somewhat among manufacturers, and the component names may not agree with Fig. 7.2(a) and (b). In the Leitz microscope in Fig. 7.1, the field diaphragm is the illumination condenser aperture, and the aperture diaphragm is the radiant field stop.

Although slightly different from Köhler illumination, Nelsonian and diffuse illumination in Fig. 7.2(a) provide similar advantages and can be used in ceramography if the microscope permits. Nelsonian illumination was devised by E.M. Nelson in the late 1800s for microscopes illuminated by a ribbon filament. Diffuse illumination is applicable to microscopes that do not have a lamp condenser. The user manuals of the better microscopes provide a procedure to attain Köhler illumination specific to each microscope model.

If the center of the specimen in the field of view does not coincide with the axis of the stage when the stage is rotated, the objective may require centering. Center the objective either with knurled thumbwheels that are concentric with the lenses or with two screws in radial directions that are 90° apart on the objective housing. Some objectives cannot be adjusted.

Magnification and Resolution

Magnification is the ratio of the size of an object in an image to the actual size of that object. Magnification is linear, as opposed to areal or spatial. An object that is 1 μm in diameter is 1000 μm, or 1 mm, in diameter in a 1000× micrograph. The projected area of that same object is magnified 1000^2, or one million times. The practical magnification by a microscope is limited by the resolution, which is limited by the wavelength of visible light. The visible wavelengths of the electromagnetic spectrum cover the range of approximately 380 to 780 nm.

Microscope magnification is dependent on several aspects of the microscope configuration. Magnification seen through the ocular is a function of the objective, body tube, and ocular. Magnification seen by the camera is a function of the objective, body tube, and camera length. Magnification seen by external attachments such as a video camera is a function of the objective and body tube again, as well as the attachment. The body tube factor is adjustable in some cases and varies among the manufacturers. Objective magnifications range from approximately 1 to 100× and body tubes from approximately 1 to 10×, and oculars are typically 10×. Microscope nomenclature varies among manufacturers—on some Zeiss microscopes, the body tube magnification is controlled by the Optovar, and the camera length is controlled by the driving drum.

The magnification factors on the microscope components are seldom exact, and so the microscope must be calibrated for each viewing mode. To calibrate a microscope, set a stage micrometer on the stage and focus on it, employing Köhler illumination. A stage micrometer is a microscope slide marked with parallel line segments of known separation width, as shown in Fig. 7.3. Stage micrometers are available for either reflected or transmitted light. Compare the reticle scale in the ocular to the stage micrometer scale to calibrate the ocular, as in Eq 1:

$$M_o = \frac{L_{SM}}{L_R} \qquad \text{(Eq 1)}$$

where, M_O is the ocular magnification as a function of objective, body tube, and ocular (μm/division); L_{SM} is the stage micrometier length (μm) corresponding to a known number of reticle divisions; and L_R is the number of reticle divisions seen in the ocular.

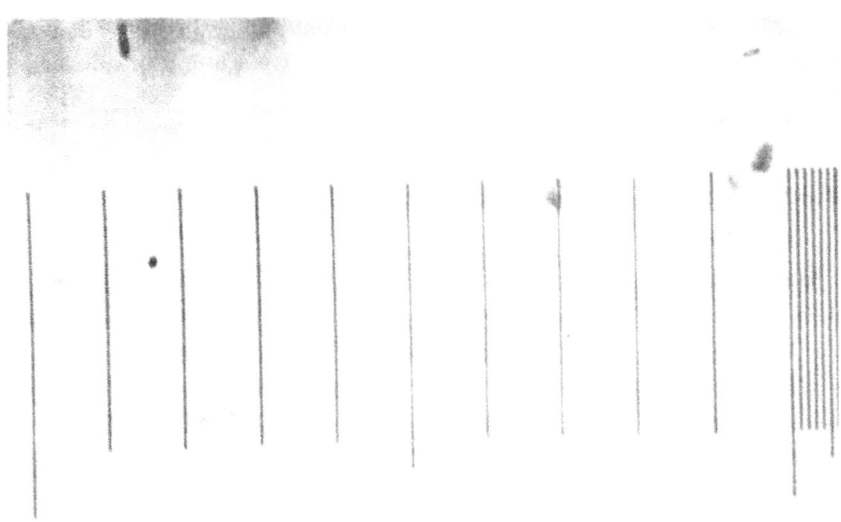

Fig. 7.3 Micrograph of a stage micrometer used to calibrate the magnification of a microscope configuration. The closely spaced line segments on the far right are 0.010 mm (10 μm) apart, and the widely spaced line segments are 0.100 mm (100 μm) apart.

To calibrate the camera, take a micrograph of the stage micrometer and note the objective, tube factor, and camera length on the back or margin of the photo. The magnification factor for the camera comes from Eq 2:

$$M_C = \frac{L_M}{L_{SM}} \quad \text{(Eq 2)}$$

where M_C is the camera magnification as a function of objective, body tube, and camera length (dimensionless); L_M is the micrograph length (mm) corresponding to a known length on the stage micrometer; and L_{SM} is the stage micrometer length (mm) seen in the micrograph.

An example of a camera calibration is shown in Fig. 7.3. Calibrate the microscope for frequently used magnifications, such as 50, 100, 200, 500, 800, 1,000, and 1500×. Some of these magnifications can be attained with more than one objective and body tube combination.

A calibration standard that can be used instead of or in addition to a stage micrometer is standard reference material (SRM) 1965, a slide containing 10 μm polystyrene spheres, available from the National Institute for Standards and Technology.

Magnification varies only slightly with position in the field of view of a microscope, as long as the entire field is in focus. Higher depth-of-field instruments, such as a scanning electron microscope, laser scan microscope, or stereomicroscope, can have considerable variation in magnification with position in the field. In these instruments, curved, tilted, or rough surfaces appear flatter than they really are, and relief is flattened into the plane of the photograph or viewing screen. For these reasons, these instruments should be used cautiously for quantitative measurements. Scanning electron microscopy as it applies to ceramography is discussed in more detail later in this chapter.

Resolution is the ability to distinguish between two discrete points, as defined in Eq 3. The resolution or resolving power is said to be high when R in Eq 3 is low. The numerical aperture term in Eq 3 is defined in Eq 4 but is usually indicated on the objective and need not be calculated:

$$R = 0.61 \frac{\lambda}{NA} \quad \text{(Eq 3)}$$

$$NA = \eta \sin \frac{\alpha}{2} \quad \text{(Eq 4)}$$

where R is the distance between two discrete points (μm); λ is the wavelength of the reflected or transmitted light (μm); NA is the numerical aperture of the objective (dimensionless); η is the refractive index (dimensionless) of the substance between the specimen or cover slip and the objective, which is 1.00 for air and 1.25 to 1.32 for most immersion oils; and α is the angular aperture, inversely related to focal length, which ranges from approximately 29° for a low-power objective to 128° for the highest-power objectives.

The resolution limit of a microscope, then, is approximately 0.2 μm under ideal conditions when violet light (λ = 380 nm) and a 100× oil-immersion objective (*NA* = 1.30) are used. This means two objects that are 0.2 μm apart can be seen as two objects rather than as one larger object. Under more realistic conditions, λ = 500 to 600 nm, and magnification greater than approximately 1300× is considered empty magnification, because resolution is at a maximum. As a rule of thumb, the useful magnification range of an objective is between 250 and 1000× its numerical aperture. The 1000×*NA* limit is disputed by some authorities. The resolution equation depends somewhat on the visual acuity of the microscopist and other factors (Ref 1).

Contrast is the relative difference of reflected signals between two discrete points, as defined in Eq 5:

$$C = \frac{S_2 - S_1}{S_2} \qquad \text{(Eq 5)}$$

where C is contrast (dimensionless); S_2 is the greater of two signals; and S_1 is the lesser of the same two signals

The two signals could be, for example, the intensities or colors reflected by adjacent objects in a microscope field.

Example 7.1: What is the magnification of the image shown in Fig. 7.3?

Solution: In the original Polaroid (Polaroid Corp.) micrograph, ten wide spaces correspond to 114 mm on a ruler. Each wide space is 0.100 mm; therefore, the ten spaces are a combined 1.000 mm, or 1000 μm, wide. From Eq 2:

$$M_c = \frac{(114 \text{ mm})}{(10)(0.100 \text{ mm})} = 114\times$$

The objective used was marked "8," and the tube factor and camera length were fixed at "12.5." One would thus expect the magnification to be close to 100×. The photograph was not necessarily reproduced in this book at exactly 1 to 1, such that the magnification in Fig. 7.3 may be somewhat higher or lower than 114×.

Depth of Field

The *depth of field* is the distance along the optical axis over which image details are in focus with acceptable clarity. The depth of field varies more or less inversely with resolution and magnification in a microscope (Ref 1). Relief polish is not recommended for microstructures with very fine features that require high magnification, because relief-polished microstructures use depth differences between adjacent features for contrast. A formula for estimation of the depth of field of a microscope is given by Eq 6:

$$d_f = \frac{\lambda\sqrt{\eta^2 - NA^2}}{NA^2} \qquad \text{(Eq 6)}$$

where d_f is the depth of field, and λ, η, and NA are the same as in Eq 3 and 4. The depth of field is large when the wavelength is large and the numerical aperture is small.

The *depth of focus,* often used interchangeably but not quite correctly with depth of field, is the range of distance along the optical axis where a sharp image can be formed on a film or retina.

Example 7.2: What is the best resolution available from a 50× dry objective viewed through a yellow-green filter? What is the depth of field?

Solution: If the aperture angle is assumed to be 128° and the index of refraction of dry air is unity, Eq 4 yields the numerical aperture:

$$NA = \eta\sin\frac{\alpha}{2} = (1.0)\sin\frac{128^\circ}{2} = 0.90$$

The yellow-green filter blocks all wavelengths except yellow-green, nominally 546 nm. The resolution is given by Eq 3:

$$R = 0.61\,\frac{\lambda}{NA} = 0.61\left(\frac{0.546\ \mu\text{m}}{0.90}\right) = 0.37\ \mu\text{m}$$

Two grains in the microscope field that are separated by a boundary region less than approximately 0.4 μm wide would be indistinguishable as separate grains. The depth of field is estimated from Eq 6:

$$d_f = \frac{\lambda\sqrt{\eta^2 - NA^2}}{NA^2} = \frac{(0.546\ \mu\text{m})\sqrt{(1.0)^2 - (0.90)^2}}{(0.90)^2} = 0.29\ \mu\text{m}$$

Any feature more than approximately 0.3 μm above or below the "plane" of focus, such as the bottom of a pore, would be out of focus.

Differential Interference Contrast

Nomarski differential interference contrast increases the contrast in some low-contrast images. Differential interference contrast uses a

Nomarski-modified Wollaston prism, contained in some objectives, and polarized bright-field light to enhance contrast on specimens that have modest amounts of relief. The polarized beam is split in two, reflected off the specimen, and recombined in the prism where interference occurs due to the phase shifting caused by path-length differences, as shown in Fig. 7.4. In Figure 8.92(b), a relief-polished spinel has a three-dimensional appearance and, in the microscope, some color contrast.

The Wollaston prism is adjusted with a lever on the objective. Differential interference contrast objectives may be marked with "POL" on the side, in addition to the magnification, numerical aperture, and working distance.

Dark-Field Illumination

The normal mode of operation of a reflected-light microscope is *bright-field,* in which the incident light is reflected by the polished surface of the specimen directly into the objective. Nearly all the visible-light micrographs in this book are bright-field images. The incident light beam is normal (at a 90° angle to the plane) or nearly normal to the polished surface, which reflects the incident beam into the objective and appears bright. Surfaces of features that are not normal to the incident beam, such as cracks, pores, and microhardness indentations, reflect the incident light away from the objective and appear dark. A *dark-field* objective increases the incident beam angle from the *z*-axis, as shown in Fig. 7.5, such that the polished surface is dark, because it reflects the incident light away from the objective, and features that have a sharp angle to the polished surface appear bright, because they reflect the incident beam into the objective.

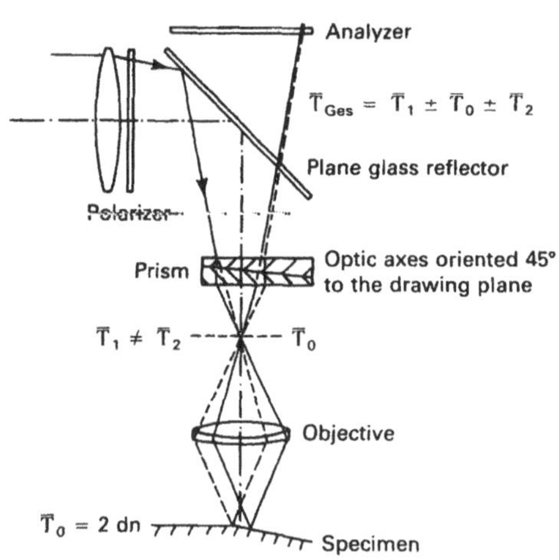

Fig. 7.4 Nomarski differential interference contrast in a reflected-light microscope. Source: Ref 1

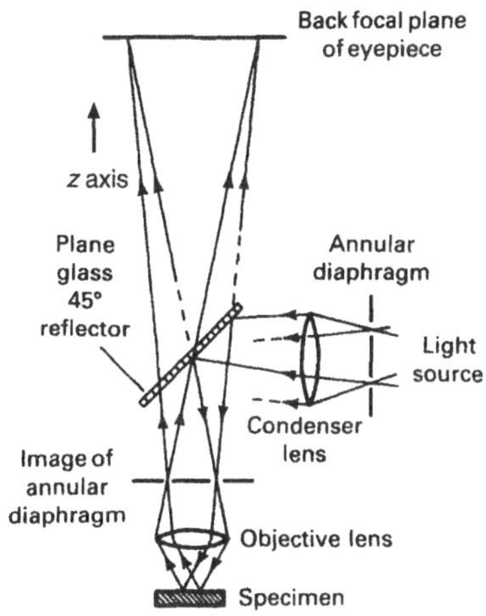

Fig. 7.5 Dark-field illumination in a reflected-light microscope. Source: Ref 3

Dark-field illumination is used to enhance microstructural features that are obscure or invisible in bright-field. The crack that is invisible in bright-field in Fig. 7.6(a) is visible in the dark-field image in Fig. 7.6(b). The large-angle incident beam in dark-field microscopy is called *oblique* illumination.

The usefulness of dark-field in ceramography is often limited by the low reflectance of ceramics. Dark-field is also applicable to transmitted-light microscopy.

Oil Immersion

Immersion oil between the objective and specimen increases the microscope resolution by increasing the numerical aperture compared to air. Oil immersion allows higher practical magnification and better resolution than dry objectives. A drop of immersion oil between and in contact with the specimen and a special, high-numerical-aperture oil-immersion objective increases the refractive index (η in Eq 4), thus increasing NA and decreasing R in Eq 3. The index of refraction for immersion oil is typically 1.25 to 1.32, and the numerical aperture of an oil-immersion lens is typically 1.30. The usefulness of oil immersion is limited in ceramography by the low reflectance of ceramics and the light absorption by the oil. The reflectivity can be improved by sputter coating. Immersion oil and oil-immersion objectives are available from most microscope manufacturers and distributors. Figure 7.7(a) shows a fine-grained alumina in oil immersion. The thermochemically etched AlN in Fig. 7.7(b) was evaporatively coated with gold prior to oil immersion. A high-index oil was used for Fig. 7.7(b), $\eta = 1.518$.

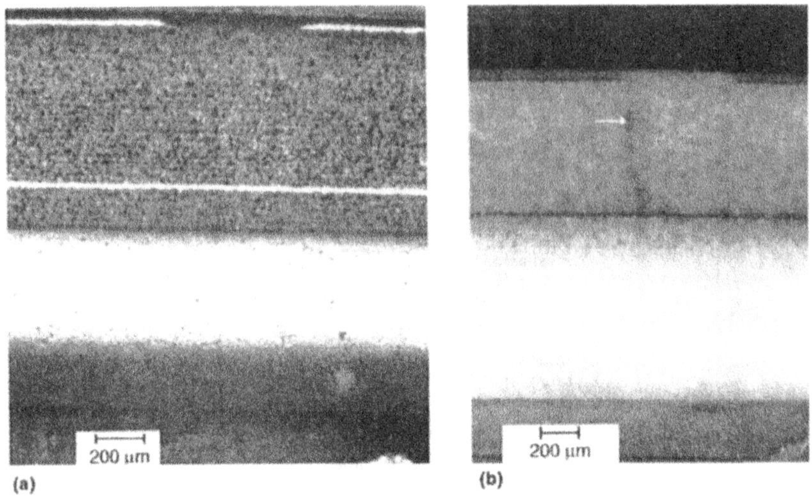

Fig. 7.6 Bright-field and dark-field image of a cross section of a multilayer interconnect board. (a) Bright-field image. The bright, thin horizontal bands are copper vias embedded in a blue dielectric layer. The alumina substrate (bottom layer) has a rounded edge due to the differential abrasion resistances of the layers. (b) Dark-field image. The vertical crack indicated by the arrow in the dielectric layer is now visible, and the copper vias are dark.

Stereomicroscopy

A *stereomicroscope* is a low-power binocular microscope that has considerable depth of field and a long working distance, due to the low numerical aperture of the objective lens. The useful magnification is limited to approximately 150×, but the depth of field is on the order of millimeters at the lowest magnifications. The stereomicroscope is especially useful for quick inspections and failure analysis. The illumination may be provided by a lamp built in to the focal tube or a fiber optic accessory for oblique illumination. The configuration allows one to hold and rotate the specimen in the hands while observing it through the microscope. A stereomicroscope is shown in Fig. 7.8.

Two types of stereomicroscope are available. The Greenough type has two nonparallel objectives, one for each eye. When one eye sees a bright, level surface, the other eye sees a darker, tilted surface. The common objective type is also binocular but has only one objective for both eyes. The image is at the same angle and in focus for both eyes simultaneously (Ref 4).

Crystal Optics

Although a detailed discussion of crystal optics is beyond the scope of this book, it is relevant to ceramography. Many ceramic crystals, especially oxides, can transmit, absorb, and refract visible light. Grain boundaries, pores, secondary phases, and lattice defects scatter transmitted light, which is why most powders and bulk ceramics appear opaque. Some key aspects of the interactions between visible light and ceramics are presented in brief in this section.

Refraction is the bending and slowing of light as it passes through a substance, quantified as a material property by the index of refraction. The symmetry of a crystal governs the number of refractive indices. The highly symmetric cubic crystals have the same refractive index along all three crystallographic axes and are said to be optically isotropic or optically homogeneous. Glass, liquids, and gases are also optically isotropic. The tetragonal and hexagonal crystals have two identical indices of refraction in the basal plane (the ω-axes) and a different one along the long or optic (ε-) axis and are said to be uniaxial. The orthorhombic, monoclinic, and triclinic crystals have three different axis lengths or angles (α, β and γ; two optic axes) and three indices of refraction and are said to be biaxial. Crystals that do not have the same index of refraction along all three axes are optically anisotropic. The *birefringence* is the difference, always positive, between the greater and lesser refractive indices in a uniaxial crystal. The birefringence of biaxial crystals is the difference between the greatest and least refractive indices. Metals and other opaque crystals do not have a refractive index.

The index of refraction (η) for monochromatic light along each crystallographic axis is defined in Eq 7. The speed of light through a crystal varies somewhat with temperature and wavelength. The wavelength dependence gives rise to the separation of colors as white light passes through a prism. The refractive index is usually measured in sodium D-light, $\lambda = 589.3$ nm, and is also related to the electrical and magnetic properties

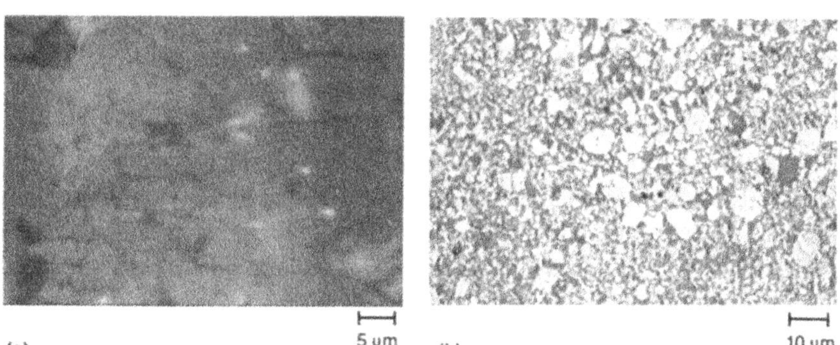

Fig. 7.7 Oil immersion. (a) 99.5% Al_2O_3 thermally etched and viewed in oil immersion with the aid of a Zeiss H-Pr reflector at high magnification. The original 2400× Polaroid photo was exposed for 60 s. (b) AlN thermochemically etched, gold plated by an evaporative coater, and viewed in oil immersion. The refractive index of the oil was $\eta = 1.518$.

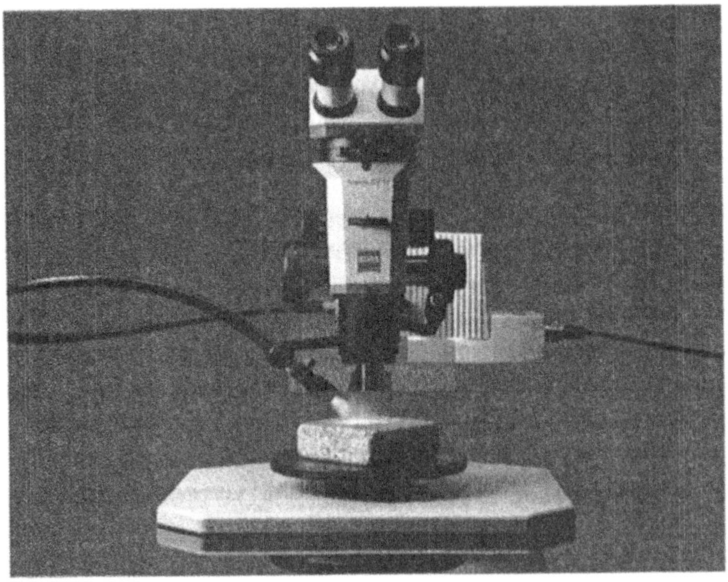

Fig. 7.8 Stereomicroscope

of the crystal. Some refractive indices for ceramics are given in Tables 7.1 and 7.2:

$$\eta_i = \frac{c}{\upsilon_i} = \sqrt{\kappa_e \kappa_m} \qquad \text{(Eq 7)}$$

where η is the refractive index along the *i*th crystallographic axis; c is the speed of light in a vacuum, 2.99792458×10^8 m/s; υ is the speed of light along the *i*th crystallographic axis; i is the ε-, ω-, α-, β-, or γ-axis of the crystal; κ_e is the dielectric constant or relative permittivity of the crystal along the *i*th crystallographic axis; and κ_m is the relative permeability of the crystal along the *i*th crystallographic axis.

When a polarized, monochromatic incident beam enters a birefringent crystal, the beam is split into two components. The birefringence is proportional to the difference in velocity between the faster beam (lower η) and the slower beam (higher η). As the two components exit the crystal and recombine, the faster component is some fraction of a wavelength ahead of, or out of phase with, the slower component. This phase shift causes constructive or destructive interference in the transmitted beam, depending on the thickness and birefringence of the crystal, the wavelength of the incident light, and the orientation of the crystal with respect to the polarizer and analyzer of the microscope. The *retardation* is the length of the phase shift, equal to the product of the thickness and birefringence.

Table 7.1 Refractive indexes of various uniaxial (hexagonal and tetragonal) ceramics at 25 °C

Ceramic	Refractive indexes η_ω	η_ε	Birefringence $\lvert\eta_\varepsilon - \eta_\omega\rvert$
SiO_2 (quartz)	1.544	1.553	0.009
BeO	1.719	1.733	0.014
α-Al_2O_3	1.768	1.760	0.008
$ZrSiO_4$ (zircon)	1.924	1.968	0.044
ZnO	2.008	2.029	0.021
TiO_2 (rutile)	2.616	2.903	0.287
α-SiC	2.654	2.697	0.043

Note: The wavelength is unspecified. Source: Ref 2

Optically anisotropic crystals that change color or intensity of color when rotated in plane-polarized white light exhibit *pleochroism.* Pleochroic crystals absorb certain wavelengths in certain crystallographic directions. *Dichroism* is pleochroism that involves only two colors.

In *chemical microscopy,* seldom used in modern ceramic engineering, the index of refraction is used to identify phases with the aid of standard refractive index liquids. Crystal optics and chemical microscopy are described in more detail in books by McCrone et al. and Insley and Fréchette.

Reflection. The reflected intensity (I_r) as a function of the normal incident intensity (I_i) at the interface of two translucent, nonferromagnetic substances is expressed by the Fresnel formula, Eq 8. The η_i terms in the equation are the refractive indexes of the two substances, such as air and polished alumina. An uncoated, polished alumina surface on a microscope stage reflects only 8% ($100 \times I_r/I_i$) of the incident light in air. In immersion oil, alumina reflects only approximately 2% of the incident light, as in Fig. 7.7. Silicon carbide reflects as much as 21% in air:

$$I_r = \left(\frac{\eta_1 - \eta_2}{\eta_1 + \eta_2}\right)^2 I_i \qquad \text{(Eq 8)}$$

Transmission and Absorption. The intensity of light transmitted through a material depends on the absorption coefficient of the material and the length of the light path. The transmitted intensity is governed by Lambert's law, Eq 9, where α is the absorption coefficient, x is the thickness of the material the light is passing through, and I_t is the transmitted intensity. The intensity of the light entering the material is I_0, that is, the incident intensity minus the reflected intensity. Either a thick section or a high absorption coefficient results in low transmission. The absorption coefficient varies with wavelength and extends both above and below the

Table 7.2 Refractive index of cubic ceramics and other optically isotropic materials

Material	Formula	Refractive index (η)
Air	. . .	1.0003
Immersion oil	. . .	1.25–1.32
Ice	$H_2O_{(s)}$	1.31
Water	$H_2O_{(l)}$	1.33
Calcium fluoride	CaF_2	1.434
Window glass	SiO_2-CaO-Na_2O	1.51–1.52
Canada balsam	. . .	1.54
Cordierite	$Mg_2Al_4Si_5O_{18}$	1.54
Spinel	$MgAl_2O_4$	1.712–1.736
Cubic zirconia	ZrO_2	2.17
Diamond	C	2.417–2.419

visible spectrum. The absorption coefficient for window glass is nearly zero for all wavelengths of visible light but large enough outside the visible range to make the glass essentially opaque to most ultraviolet radiation. In a polycrystalline material, the absorption coefficient is highly dependent on the microstructure, because grain boundaries, pores, and other microstructural features absorb and scatter transmitted light:

$$I_t = I_0 e^{-ax} \qquad \text{(Eq 9)}$$

Petrography—Transmitted Light and Thin Sections

Ceramics that are not totally opaque can be thinned and observed in transmitted light. Transmitted light is especially effective for the study of optically heterogeneous (noncubic) ceramics, such as alumina, in which the refractive index varies with crystallographic direction. Geologists often use this method to study mineral agglomerates; hence, it is also called petrography.

To view a ceramic in transmitted light, prepare a thin section, as described in Chapter 6. Configure the microscope for transmitted light and Köhler illumination. Transmitted-light microscopy requires a substage condenser and a substage polarizer. Set the slide on the stage of the microscope, focus on the specimen, and cross the polars by rotating the analyzer angle to maximize contrast. Figures 7.9(a–d) demonstrate the principles of polarized light in a microscope.

Crossed polars normally block the light beam. The polarizer allows only one orientation to pass, as in Fig. 7.9(a), and the analyzer blocks that orientation in Fig. 7.9(b) and (c), because its permissible direction is perpendicular to the polarizer. Translucent noncubic crystals split the transmitted light into two components corresponding to the dissimilar crystallographic axes, as in Fig. 7.9(d). The double-refracted, phase-shifted beam leaving the crystal is reoriented with respect to the polarizer direction when it reaches the analyzer. The reoriented beam is partially transmitted

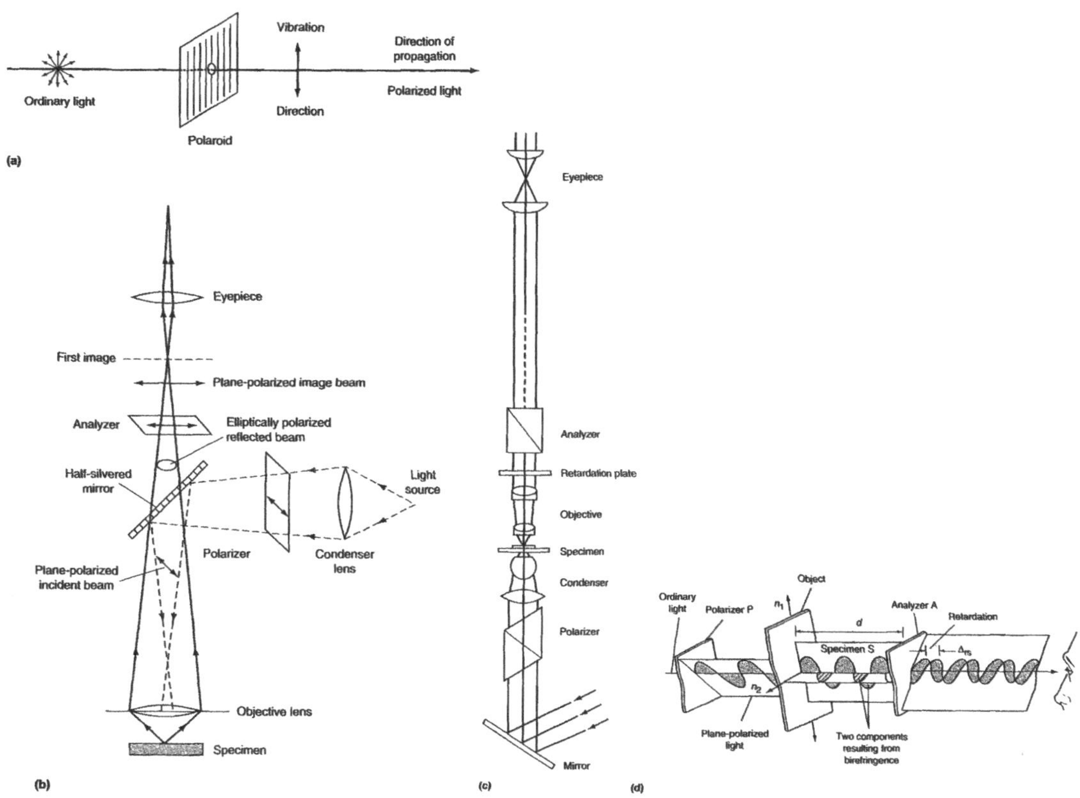

Fig. 7.9 Polarized light. (a) Polarization of light by a polarizing filter. Source: Ref 2. (b) Polarized reflected light in a microscope. Source: Ref 3. The polars are crossed when the transmission direction of the polarizer is perpendicular to that of the analyzer. (c) Polarized transmitted light in a microscope. Source: Ref 5. (d) Transmitted light split into two components. Source: Ref 4

by the analyzer onto the oculars. As the microscope stage is rotated 360°, the light transmitted by each noncubic grain changes from approximately zero to some maximum intensity four times. The angles of zero transmission, when the grain is dark, are the extinction angles. The intermediate intensities are the result of varying degrees of interference of refracted wavelengths. In a randomly oriented ceramic, some grains are at the extinction angle, some grains are at maximum transmission, and most grains are somewhere between the two extremes, resulting in contrast. Cubic ceramics and the glass slide cannot double-refract the transmitted light, because their indices of refraction are the same in all directions. Cubic ceramics and glass appear dark at all stage angles when viewed through crossed polars.

The thin-section method is fast and easy compared to the thick-section methods but is only semiquantitative in most cases, due to the overlapping and stacking of grains superimposed on each other in the image plane. Some grains may be viewed in their entirety, and grain sizes tend to appear slightly larger in the thin section for this reason. One way to use thin sections to evaluate grain size is by making thin sections from ceramics of known composition that have been measured for grain size by any of the methods discussed in Chapter 9 and comparing other thin sections to the standards. For best results, only compare test specimens to standards that are of similar composition, such as a 94% alumina to another 94% alumina. Determine the smallest and largest grain sizes directly in the microscope reticle or by measuring a micrograph.

Images of thin sections, especially Polaroid micrographs, tend to contain a wide range of transmitted intensities. Some grains in the micrograph are overexposed while others are underexposed when the majority are adequately exposed. High color contrast is obtainable when the thin section is viewed through a full-wavelength compensator. The retardation plate shown in Fig. 7.9(c) could be a full-wavelength or quarter-wavelength compensator. Preferred orientation (discussed in Chapter 10) is indicated when the extinction (darkening) angle is the same for a majority of grains as the stage is rotated.

Optically homogeneous (cubic) ceramics, which have no contrast between grains in polarized transmitted light, can be measured in thin sections when the contrast is high between grains and grain-boundary phases. Figure 8.94 shows a partially stabilized zirconia through uncrossed polars where the cubic primary grains all have essentially the same refractive index in all orientations. The dark grain boundaries are marked by noncubic secondary phases, namely monoclinic or tetragonal zirconia precipitates.

Some ASTM procedures that apply to thin sections, including C 856 and E 211, are listed in Appendix A.

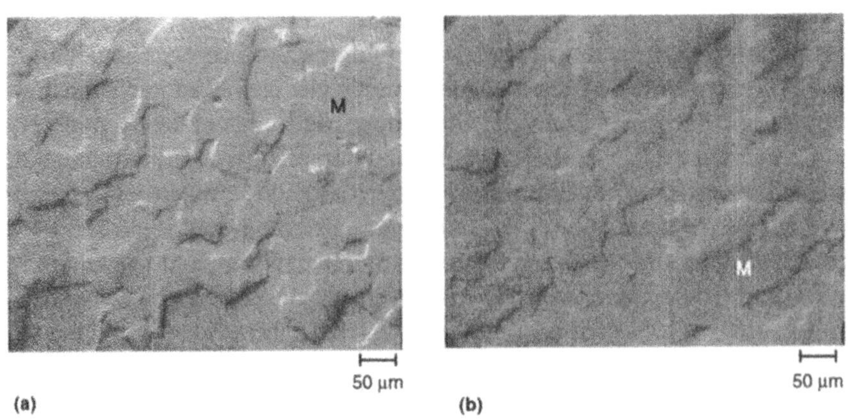

Fig. 7.10 Replication. (a) $MgAl_2O_4$ spinel original relief polished with 0.05 μm γ-Al_2O_3 and viewed in differential interference contrast. The "M" marks the same grain in both images. (b) Replica of $MgAl_2O_4$ spinel at left on 129 μm acetate tape. The replicated side of the tape is up, away from the glass slide. The image is a negative, or mirror reflection, of the original. The "M" marks the same grain in both images. Viewed in differential interference contrast

Replication and Field Ceramography

Ceramographic replication is the process of making copies—replicas—of microstructures that do not alter the original. The replica is not just an image; it is a 1 to 1 scale model complete with texture. It is important to note that the replica is a mirror-image model or negative of the original. That is, ridges on the original correspond to valleys on the replica, and left on the original is right on the replica. Replication can be easily done in remote locations or in situ; thus, it is also called field ceramography or in situ ceramography.

Replication, in and of itself, is nondestructive, but the preparation of the original is often a destructive test. Several metallographic equipment companies make portable grinding and etching kits that are used primarily in power plants and petrochemical refineries to detect the onset of creep cavitation and other catastrophic failures in pipelines, tanks, and valves. These replication kits can also be applied to ceramic devices.

Three-dimensional objects, such as fracture or curved surfaces, are reduced to two dimensions by replication, thus allowing one to use a microscope to see the fine detail of a fracture. Replication is fast and easy compared to the mount-polish-etch methods of Chapters 3 to 5 and nearly as sensitive. An original and replica are shown side by side in Fig. 7.10(a) and (b). Mounted and polished specimens can be replicated so that the as-polished surface is preserved in a replica. Other materials besides acetate tape can be used for replication, such as the liquid mounting resins of Chapter 3 used with liquid mold release on the surface to be replicated.

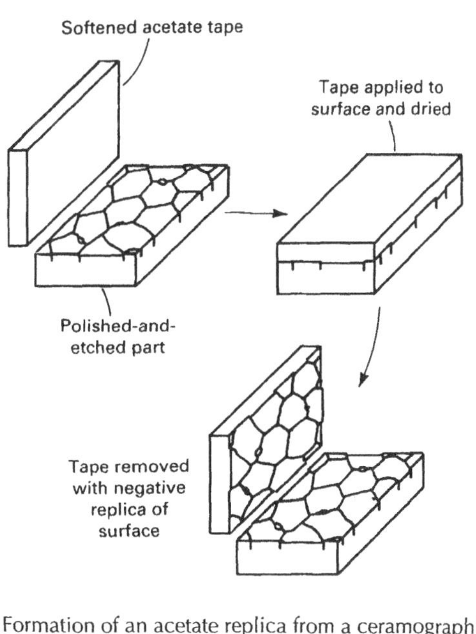

Fig. 7.11 Formation of an acetate replica from a ceramographic or metallographic section. Source: Ref 7

The negative replica can be replicated to create a positive replica, although some detail may be lost with each replication. An example of this is a carbon replica, formed by evaporative coating, of an acetate replica, formed by the subsequent procedure. The acetate replica can be sputter coated for scanning electron microscopy (SEM) observation, as described in the section "Sputter Coating" in this chapter. Carbon replicas are conductive and often used in electron microscopy.

Radioactive ceramics, such as many of the rare earth oxides (UO_2, ThO_2, etc.), can be replicated (Ref 6). The replica itself is not radioactive and can be handled without any concern for radiation safety. Similarly, replicas of other hazardous materials can be used for imaging and measurement while the source is safely isolated.

An ASTM procedure, E 1352, is listed in Appendix A. A paper by Dillinger describes metallographic replication.

Replication Procedure. The steps are:

1. Cut a piece of 129 μm (0.005 in.) cellulose acetate replication tape large enough to cover the area of interest and fit on a microscope slide, typically a 10 mm square.
2. Soak the tape in acetone for 10 to 20 s or until it softens to a gelatinous state and lay it on the area to be replicated, as shown in Fig. 7.11.
3. While the acetone is evaporating, prepare a microscope slide by putting a piece of double-stick adhesive tape (e.g., 3M Scotch 400, 665, or 136), about the same size as the acetate, in the center of one side of the slide. Peel the backing off the adhesive tape.
4. Use tweezers to carefully remove the acetate from the original specimen once the acetate has dried. Gently press the acetate onto the adhesive tape with the replicated side of the acetate up, away from the adhesive.
5. Darken the backside of the slide with a felt-tip marking pen or electrician's tape or cover it with foil to improve reflectance. Sputter coat the slide with metal particles (see the section "Sputter Coating" in this Chapter) for SEM observation or for improved visible-light reflectance. Tilt the slide in the sputter coater to create a shadow effect between neighboring features of different heights.
6. Observe the replica under reflected light or in the SEM. Note that the replica is a mirror image, or negative, of the original. A replica of the first replica is a positive.
7. A replica can also be made with castable resins, similar to the procedures in Chapter 3. Spray a thin coat of aerosol mold release on the surface to be replicated and the mold cup or ring and any other part of the specimen that is likely to come in contact with the resin. Tape the ring to the specimen surface, if the specimen is large enough, or else make a device that can suspend the specimen atop the cup. Prepare the

resin and pour it into the ring or cup. With the cup, suspend the specimen so that the surface of interest is shallowly submerged in the resin, but the rest of the specimen is not wetted by the resin. A paper clip or alligator clip hanging from a wire on a chemist's ring stand can be used to suspend the specimen. Allow the resin time to cure undisturbed. When the resin has hardened, gently peel the ring from the ceramic or the ceramic from the cup. The resin surface is the replica.

8. Some metallographic manufacturers offer a replication kit that uses two-component silicone rubber. The silicone is squeezed from a tube, much like a caulk gun, as a viscous liquid onto the surface to be replicated. The two components are mixed in the application and react to form a polymerized rubber replica.

Sputter Coating

Although cathode sputter coating is used primarily to make ceramics conductive for SEM purposes, it can also improve images of ceramographic sections in the microscope by increasing the reflectance, as demonstrated in Fig. 7.12(a) and (b). A conductive layer, 10 to 100 nm thick, is applied in a vacuum chamber backfilled with argon in a high electric field for 5 to 10 min. The conductive coating may be gold, palladium, platinum, silver, copper, aluminum, or carbon (Ref 8). Gold-palladium alloy condenses uniformly and has good adhesion to most ceramics. After a few minutes of cooling, the coated specimen is ready for viewing in the microscope. Carbon evaporative coating is less beneficial for microscope viewing but useful for x-ray microanalysis in the SEM. The specimen can

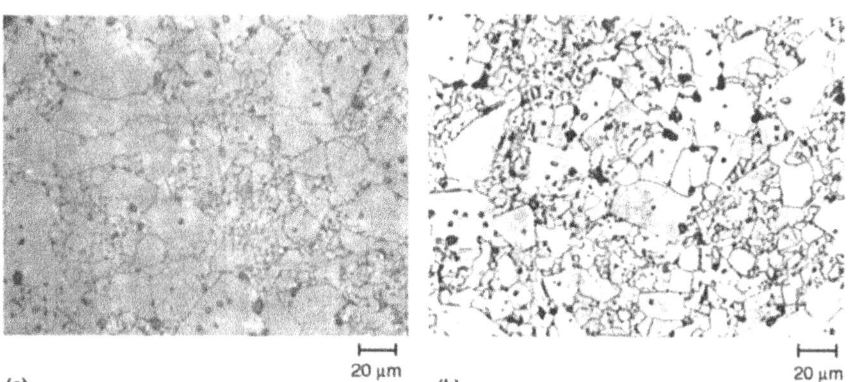

Fig. 7.12 Sputter coating. (a) 99.5% Al_2O_3 etched in boiling phosphoric acid for 30 s. Polaroid photo with 15 s exposure. The average grain size is 8.7 μm by the circle-intercept method. (b) The same 99.5% Al_2O_3 after the surface was sputter coated with a gold-palladium alloy. Polaroid photo with 10 s exposure.

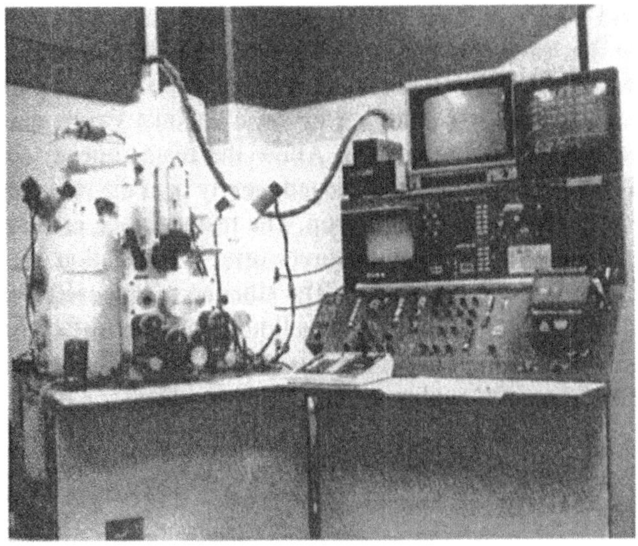

(a)

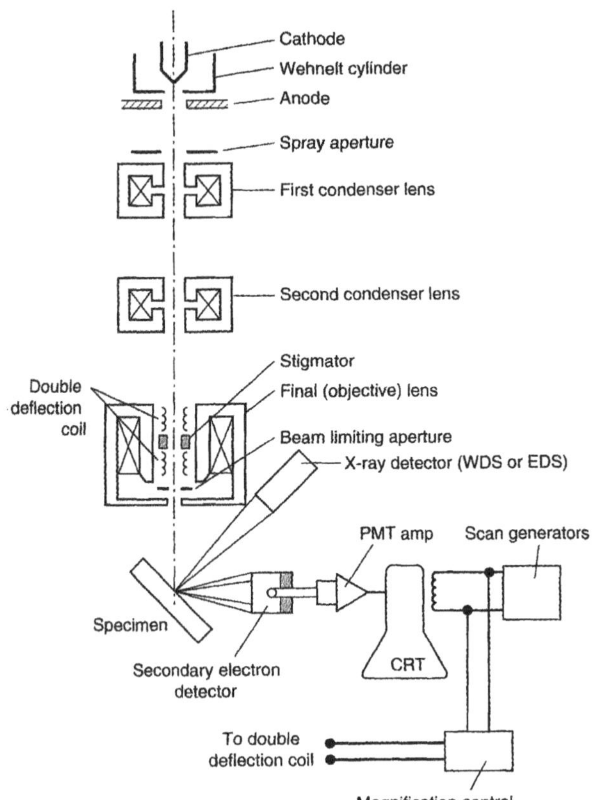

(b)

Fig. 7.13 Scanning electron microscopy. (a) Scanning electron microscope and x-ray microanalyzer. (b) Schematic of the electron and x-ray optics of a scanning electron microscope and electron probe microanalyzer WDS, wavelength-dispersive spectroscopy; EDS, energy-dispersive spectroscopy; PMT, photomultiplier tube; AMP, amplifier; CRT, cathode-ray tube. Source: Ref 9

be tilted in the sputter coater to achieve a shadowing effect on topographical features. Sputter coater manufacturers and distributors include Denton Vacuum, Fullam, Ladd, Pelco, and Polaron.

Metallic sputter coating is similar to the interference layer etching described in Chapter 5, except that the metallic coating is opaque, conductive, and highly reflective. The interference layer is necessarily translucent, usually nonconductive, and reflects only a small amount of the incident light. A metallic sputter coating opacifies the surface of a translucent ceramic and reflects the incident light almost as well as a mirror.

Do not sputter coat a specimen prior to etching or microindentation hardness testing. The sputter coating can be removed with light relief polish (see Chapters 4 and 5), although some coating will remain in the pores, grain boundaries, and microcracks. Sputter coating is also useful in the preparation of ceramic pellets for dielectric testing.

Scanning Electron Microscopy

The SEM constructs an image from electronic signals returned to its detectors by a specimen that has been bombarded by an incident electron beam in a vacuum chamber. The signals returned by the specimen include backscattered (reflected) electrons, secondary (ejected) electrons, Auger (pronounced like "OJ" in French) electrons, and electromagnetic radiation from infrared to x-rays. The SEM has much greater practical magnification, resolution, and depth of field than the microscope and can be used to view microstructural features that are much too small to be resolved in the visible spectrum. The SEM configuration is shown in Fig. 7.13(a) and schematically in Fig. 7.13(b).

The electronic signals are sensitive, in varying degrees, to the elements in the specimen and to the topography or relief of the specimen surface but not to colors. The backscattered electrons are more or less analogous to reflected light in the microscope. The secondary and Auger electrons are ejected by the specimen in inelastic collisions and are somewhat analogous to fluorescence. The electronic and x-ray signals from the specimen originate not only from the surface but include interactions from a finite volume beneath the surface as well (Ref 10). The x-rays have the broadest and deepest region of origin within the specimen. The backscattered electrons originate from a zone of intermediate depth and breadth, and the contrast is primarily due to differences in atomic number and crystallographic orientation. The secondary electrons are emitted from a shallow, narrow region starting at the specimen surface. The contrast in secondary electrons is primarily due to topography and the edges of pores and grains. The contrast is seen as brightness and gray level in the micrograph or viewing screen of the SEM. Other signals of varying degrees of importance are also generated by the electron beam/specimen interaction, including emission of visible light. The visible light emission, *cathodoluminescence,* can provide color contrast.

The large depth of field of the SEM has both advantages and disadvantages in ceramography. The SEM can have a substantially large three-dimensional object in focus all at one time, but the magnification varies throughout the field of view. Loose ceramic powder can be viewed by SEM, as shown in Fig. 7.14, and the particle size can be visually estimated. The SEM is very useful for fractography and other unpolished surfaces but should be used sparingly for quantitative measurements, such as grain size. The SEM should be considered a supplement to, rather than a replacement for, the microscope.

The SEM stage, which is inside a vacuum chamber, can be rotated and tilted. Stereoscopic pairs (discussed later in this Chapter), two micrographs of the same object taken at angles 5 to 10 ° apart from each other, are used for three-dimensional imaging. This feature is very useful in fractography and morphological studies of microstructural features. Because of the vacuum, volatile substances and outgassing specimens should not be used in the SEM, such as specimens mounted in castable resins.

Nonconductive ceramics must be sputter coated with a layer of metal or carbon before they can be viewed by the SEM. Electrical conductivity through the encapsulation is maintained by either a conductive encapsulation (described in Chapter 3) or a stripe of conductive paint between the

Fig. 7.14 96% Al_2O_3 powder prior to compaction and sintering (Ref 11). Surface area = 0.6 m^2/g, according to the manufacturer. The powder was sprinkled onto a carbon-painted metal SEM stud and gold sputter coated.

specimen and the specimen holder set screw. The conductive paint, graphite or metallic particles suspended in amyl acetate, is also useful for attaching an unmounted specimen to a metal SEM stub. Conductive double-stick tape is also available for attaching the specimen to the stub.

Scanning electron microscope capabilities can be extended with the addition of an electron probe x-ray microanalyzer for microchemical analysis. The two types of microanalyzer are the energy-dispersive (EDS or EDX) and wavelength-dispersive (WDS) x-ray spectrometer. The WDS is ten times as sensitive as the EDS, but the EDS is faster. The WDS can detect as little as 0.01 mass% (100 ppm) of an element, and the EDS can detect as little as 0.1 mass% (1000 ppm). The spectrometers detect and analyze the characteristic x-rays emitted by the specimen as a result of electron bombardment to semiquantitatively determine the elements—not compounds—present, as in Fig. 7.15. A short working distance, say, 25 mm or less, depending on the brand of SEM, is needed for optimal accuracy and sensitivity. Elements of low atomic number, for example, $Z \leq 10$ (neon), may not be detected accurately, due to absorption of x-rays and

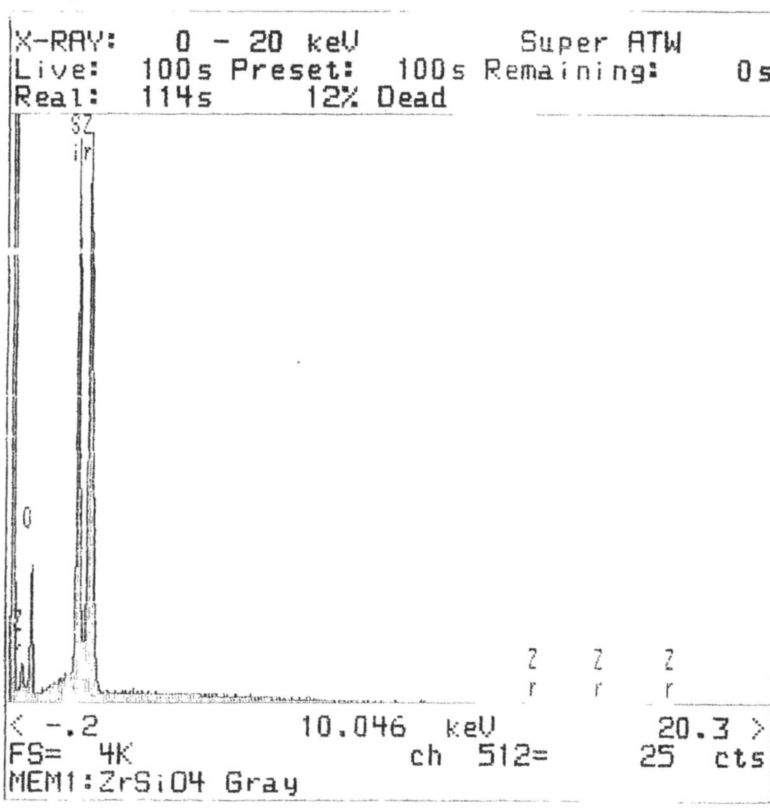

Fig. 7.15 Wavelength-dispersive spectroscopy spectrograph of the zircon ($ZrSiO_4$) grinding pellet shown in Fig. 8.79. The analysis indicates 71 mol% O, 14% Si, and 15% Zr. The initial peak on the left edge is noise.

detector noise. The spectrometer can generate an x-ray map of the specimen, that is, an image that shows in pseudocolor the locations of a particular element, as in Fig. 7.16. The spatial resolution is limited to approximately 1 μm. The sputter coating is detected along with the specimen, and the choice of carbon or metal should be considered accordingly (Ref 13).

Kikuchi diffraction is a recent addition to the SEM capabilities. The Kikuchi diffractometer, a hardware accessory on the SEM, acquires and analyzes a diffraction pattern from backscattered electrons. The system computer uses the Bragg angles and lattice spacing from the electron backscatter diffraction pattern, elemental analysis from the WDS or EDS, and a database of diffraction patterns to identify compounds and crystal structures.

Figure 7.17 modestly demonstrates the magnification and resolution capabilities of the SEM. Scanning electron microscope resolution is also governed by Eq 3 and 4. The angular aperture of the SEM is very small, less than one degree. The wavelength of an electron is several orders of magnitude smaller than that of visible light, less than 0.004 nm in a 100 kV electric field. Thus, the resolution of the SEM, with $\alpha = 0.15°$ and $\lambda = 0.00370$ nm, is approximately 0.9 nm.

Ceramics that cannot be etched, or for which no etchant is known, can be microstructurally characterized by SEM in many circumstances. The grain size of these ceramics is estimated from a reasonably flat, mostly transgranular fracture surface micrographed by SEM.

Some ASTM procedures that apply to SEM are listed in Appendix A; these include E 748, E 766, and E 986. Scanning electron microscopy reference texts include books by Goldstein et al., Rochow and Tucker, and Murr.

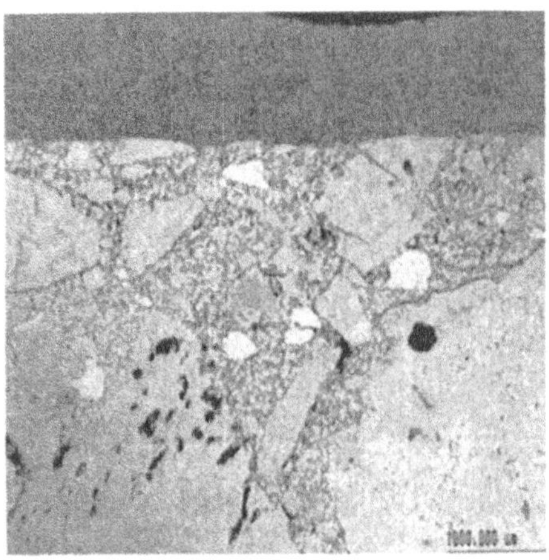

Fig. 7.16 X-ray map of slag-corroded Cr_2O_3-Al_2O_3 refractory brick from a gasifier. Silicon is shown in red, iron in green, and calcium in blue (Ref 12). SEM photo by W.K. Collins. For color version of Fig. 7.16, see endsheets of book.

Other Microscope Types

Several other types of electron microscope besides the SEM are available. Most are used primarily for research rather than routine inspection. The transmission electron microscope (TEM) generates an image from an electron beam passing through a specimen that has been thinned by an ion mill or electrojet or a carbon replica. The specimen must be thinned to somewhere between 10 nm and 1 μm to transmit the electron beam. The TEM can also generate an electron diffraction pattern useful for identifying the compound or crystal structure of a single grain. The resolution of a TEM approaches 0.4 nm, suitable for viewing dislocations and stacking faults in a crystal lattice (Ref 14). Ion mills are mentioned in the section "Other Etching Methods" in Chapter 5, under the heading "Ion Etching." An electrojet polishes and thins a positively charged specimen by pumping a stream of anionic solution against it. The ion mill or electrojet may be used to thin specimens for transmitted visible-light microscopy as well as TEM. A TEM image of alumina, with secondary phases identified, is shown in Fig. 7.18.

The analytical electron microscope (AEM) combines a TEM with x-ray microanalysis and microdiffraction into an instrument that can simultane-

Fig. 7.17 Tape-cast 96% Al_2O_3 substrate chip-mounting surface, not polished or etched. The average grain size is approximately 0.9 μm by the circle-intercept method, too small for the microscope even if the surface had been polished. The caption indicates an arbitrary photo No. 0010, and the electron beam was accelerated in a 20 kV electric field. The original Polaroid photo magnification was 4000×, and the working distance was 26 mm.

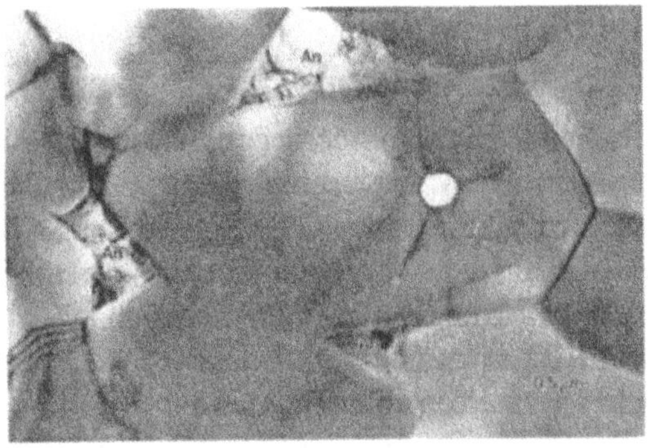

Fig. 7.18 Transmission electron microscopy image of 96% Al_2O_3 with anorthite ("An") as an intergranular phase. The white hexagon in the largest grain is a pore. Courtesy of C.P. Doğan

ously provide images of very fine structures, chemical information, and a diffraction pattern. When the electron beam of the AEM is focused to a fine probe and scanned (rastered) over the surface of an ultrathin specimen, the AEM becomes a scanning transmission electron microscope (Ref 15).

The scanning tunneling microscope (STM) holds a conductive probe approximately 1 nm from the surface of an electrically conductive solid while scanning the surface. The probe does not touch the surface. The current from electrons "tunneling" between the probe and surface electron clouds is analyzed and used to adjust the probe distance so as to maintain a constant signal. The signal and the *x-y* position of the probe are used to generate a contour map of the surface. The STM, invented circa 1981 by Rohrer and Binnig at the IBM Zurich Research Laboratory, can resolve atoms (Ref 4).

The atomic force microscope (AFM) is similar to the STM and also analogous to a phonograph on a much, much smaller scale. The AFM scans a solid surface with a ceramic or semiconductor tip mounted on a cantilever spring. The tip is in contact with the surface, and the surface does not have to conduct electricity. The attraction or repulsion forces between the tip and the surface deflect the spring, very similar to the piezoelectric stylus of a phonograph on the surface of a vinyl record. An angled laser beam detects the deflection of the spring. The magnitude of the force and the *x-y* position of the tip are used to generate a contour map (Ref 15). The AFM is one type of scanning probe microscope. Scanning probe microscopes are available from Digital Instruments, Quesant Instrument Corp., and TopoMetrix, all in California.

Some other imaging instruments and techniques not described here are rare in the ceramics industry but have niche applications in research and development. These include field emission microscopy, field ion microscopy, hot-stage microscopy, cold-stage (cryogenic) microscopy, x-ray microscopy, infrared microscopy, and scanning near-field light microscopy. These are all mentioned briefly in Rochow and Tucker's book.

Transmission electron microscopy textbooks include those by Williams and Carter, Von Heimendahl, and Murr. A book of proceedings from a STM and AFM conference is available from Kluwer. Rochow and Tucker's book describes electron microscopes along with other optical instruments and techniques.

As-Fired Surface

Some ceramics can be evaluated for quality-control purposes with reasonable accuracy from micrographs of their as-fired surface. This method is fast and easy compared to the mount-polish-etch method but not as accurate or precise. Best suited for this method are ceramics that have a smooth, flat, unglazed surface, such as the alumina substrate shown in Fig. 7.19.

As-fired surfaces, because of their relative roughness compared to a polished surface, must be viewed under a high depth-of-field instrument, such as a SEM, laser scan microscope, or stereomicroscope, for high, medium, and low magnification, respectively. Grain size and other measurements should be considered approximate due to depth distortion in the focal "plane."

Stereo Pairs

Stereo pairs are two micrographs of the same subject, taken from positions 5 to 10° apart, and viewed side by side in a stereoscope. The two

Fig. 7.19 Scanning electron microscopy image of a tape-cast 96% Al_2O_3 substrate chip-mounting surface as-fired (not polished or etched). The "peanut brittle" texture just to the right of the center of the photo may be due to a vitreous phase.

photos in a stereo pair appear as one with a third dimension, depth. Stereo pairs are used in many professions, but in ceramography, they provide a more detailed view of three-dimensional structures, such as fracture surfaces and voids, than single, two-dimensional micrographs. The SEM is frequently used to make stereo pairs, because its stage can be tilted easily and precisely, and it has a large depth of field. Figs. 7.20(a) and (b) show two SEM micrographs that look virtually the same, except they are tilted from each other by a few degrees. Observe these photos with a stereoscope to see the depth of the fracture features.

A paper by Dillinger offers another perspective on stereo pairs.

Acoustic Microscopy

High-frequency sonic waves, rather than visible light or an electron beam, are used to generate microscopic images in *acoustic microscopy.* Acoustic microscopy is more of a specialized nondestructive testing (NDT) technique than a means of microstructural analysis but has some specialty applications in ceramography. The three distinct types of acoustic microscopy are scanning acoustic microscopy, scanning laser acoustic microscopy, and C-mode scanning acoustic microscopy. Each type has specific applications that are not well suited to the other two (Ref 16).

The *scanning acoustic microscope* (SAM) combines the immersion mode of ultrasonic NDT with an acoustic magnifying lens. The image is constructed from acoustic signals reflected through a liquid couplant by the surface and shallow subsurface features of the specimen. The motorized sensing unit scans the specimen in an *x-y* pattern to create images in three dimensions. The SAM uses only Rayleigh (surface) waves and has

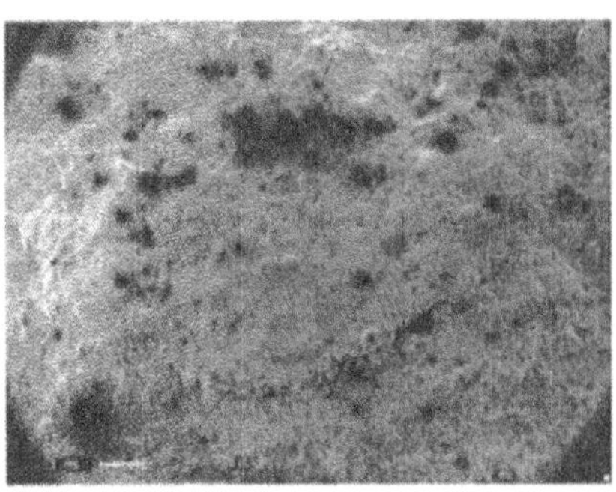

(a)

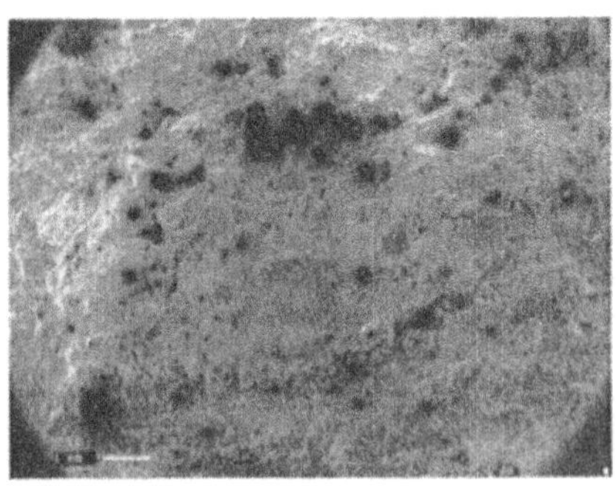

(b)

Fig. 7.20 Stereo pair. (a) SEM photo of the fracture surface of a SiC heating element that failed after numerous furnace cycles. An electric arc at the time and location of the final crack progress melted the region at the center of the image. (b) Same specimen and location as (a) at left but viewed with the SEM stage tilted 10°. Look at the two images with a stereoscopic viewer to see the depth dimension.

the highest resolution of the three types of acoustic microscopy but very little depth of penetration. The SAM is used primarily for detection of surface and near-surface defects in microelectronic devices. A SAM is shown in Fig. 7.21(a) and as a block diagram in Fig. 7.21(b).

The frequency range of the SAM acoustic signal is approximately 1 MHz to 10 GHz. The resolution in Eq 10 improves, but the depth of penetration decreases with increasing frequency. A typical lens aperture angle is 5 to 10°. Equation 10 is similar to Eq 3 and 4. Frequencies as high as 5 GHz can be used in water to achieve resolution comparable to a microscope, approximately 0.3 μm. Frequencies as high as 16 GHz can be used in liquid helium to achieve resolution on the order of 15 nm:

$$\varpi = \frac{0.61\lambda_0}{\sin\theta_0} = \frac{0.61\upsilon_0}{f\sin\theta_0} \qquad \text{(Eq 10)}$$

where ϖ is the acoustic resolution (μm); λ_0 is the wavelength (μm) of the acoustic signal, equal to velocity divided by frequency (v_0/f); θ is the angle of the lens aperture; $\sin\theta_0$ is the numerical aperture; v_0 is the velocity (m/s) of sound in the acoustic couplant, for example, water, and f is the frequency (MHz) of the ultrasonic beam.

The *scanning laser acoustic microscope* (SLAM) detects transmitted waves from a transducer beneath the submerged specimen as surface ripples by a scanning laser above the specimen. A submerged coverslip is placed on the surface if the specimen is rough or irregular. The SLAM

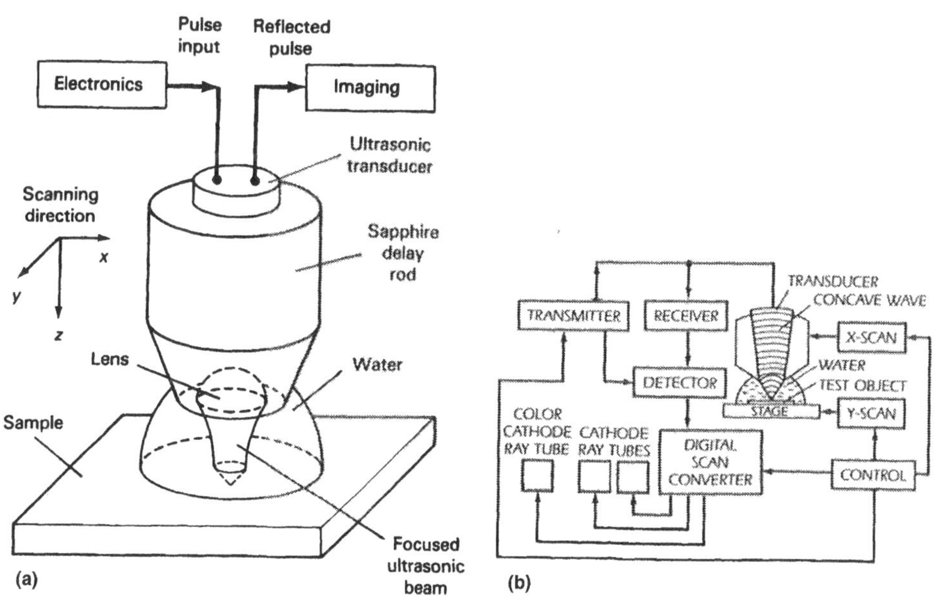

Fig. 7.21 Scanning acoustic microscope. (a) Reflective-type scanning acoustic microscope. (b) Block diagram of a reflective-type scanning acoustic microscope (Ref 16). The beam angle is measured from the normal to the specimen surface.

operates in real-time and can be used to observe the propagation of a fracture while the specimen is in tension. The frequency range for a SLAM is typically 10 to 500 MHz.

The *C-mode scanning acoustic microscope* (C-SAM) is very similar to the SAM but uses a lower-frequency ultrasonic beam of 10 to 100 MHz at a lower beam angle. With the lower frequencies and angles, the C-SAM can generate an image at any depth in the specimen. The depth of penetration is limited by the attenuation (damping and scattering) of the ultrasonic beam by the material, the focal length of the lens, and refraction of the beam. The image in a C-SAM is a gray-scale C-scan of reflected (pulse-echo) signals. The three types of acoustic microscope are compared in Fig. 7.22.

In acoustic microscopy, the specimen is immersed in a liquid bath along with a ceramic piezoelectric transducer. The liquid is usually water, but other couplants may be used. Acoustic energy—sonic waves—does not pass easily between substances that have a large mismatch in acoustic impedance. The difference in acoustic impedance between gases and solids is much greater than between liquids and solids, which is why the specimen and transducer are immersed in a liquid. The transducer vibrates by straining cyclically in an alternating electric field and emits a high-frequency acoustic pulse. Most of the pulse is reflected at the specimen-water interface, that is, the specimen surface. Some of the pulse—depending on the type of acoustic microscope and the angle of the beam—is transmitted through the first interface and reflected or scattered by the next interface, which could be a pore, grain boundary, crack, or back surface. The acoustic signal is also refracted, or changes speed and direction by some angle, as it passes through each interface. The SAM uses a large beam angle to generate surface waves that penetrate only to a depth of approximately one wavelength. The reflected signals can be scattered in every direction, depending on the nature of the interfaces, but some are reflected back to the transducer through an acoustic lens. The transducer is vibrated by the reflected signals and generates an electric signal that can be amplified and converted into images. The portion of the signal that is not reflected to the transducer, that is, scattered and absorbed, is *attenuated* by the specimen.

The three dimensions detected by acoustic microscopes are not presented in one image but in three views. The A-scan is a z-t plot of pulse amplitude versus time and can be used to measure the thickness of the specimen. The B-scan uses parallel sets of A-scans to present the subsurface feature distribution through a cross section of the specimen thickness, a z-x or y-z plot. The C-scan or plan view, the most common, provides an x-y distribution of the size and position of subsurface features but not depth.

Piezoelectric zinc oxide (ZnO) is used as the ultrasonic transducer in acoustic microscopes. The density of ZnO is 5.68 g/cm^3 and its sonic ve-

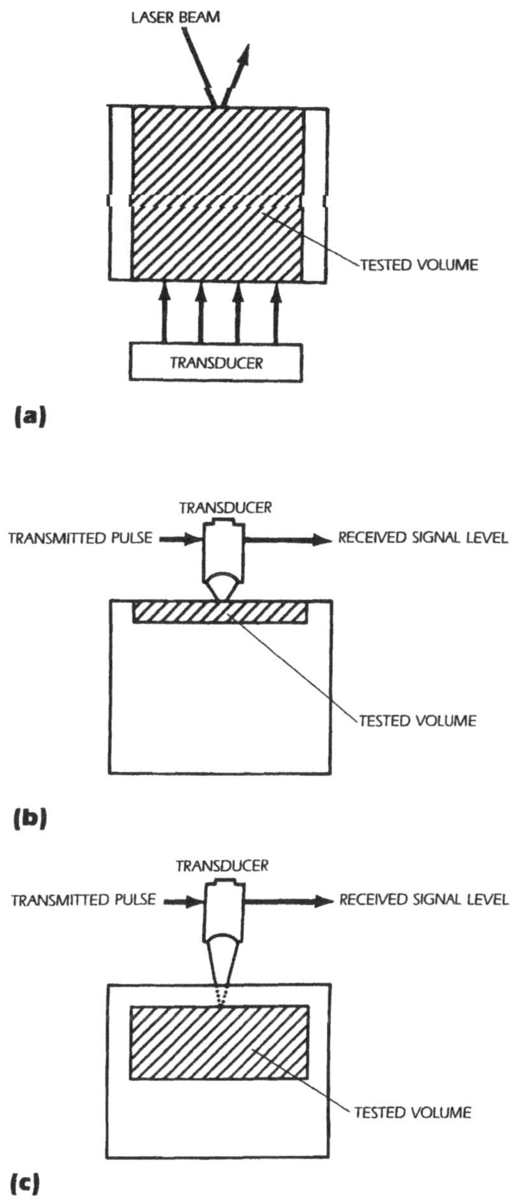

Fig. 7.22 Comparison of the three types of acoustic microscope. The zones of application are shown as cross-hatched areas within the test object. (a) SLAM. (b) SAM. (c) C-SAM. Source: Ref 16

locity is 6330 m/s. The ultrasonic transducer is both an emitter and a receiver of acoustic signals. Transducers use a material property to convert one form of energy or signal into another in a reversible reaction. Piezoelectric transducers convert electrical energy into mechanical energy, and vice versa. *Piezoelectric* or *electrostrictive* ceramics strain in the presence of an electric field or generate an electric field while straining under a mechanical stress.

Some acoustic properties of ceramics and other materials are given in Table 7.3. The difference in acoustic impedance—the product of density and sonic velocity—between alumina and air is some five orders of magnitude, whereas the difference between alumina and water is only one order of magnitude. The velocity of sound through alumina is approximately thirty times faster than sound in air. The crystal structure has an effect on acoustic properties, evident from the difference between 1020 steel, where the iron is body-centered cubic, and 302 stainless steel, where the iron is face-centered cubic. The differences between liquid water and ice are also significant. Although not indicated by the table, the acoustic properties are affected somewhat by the grain size and porosity (Ref 18). The acoustic property relationships include Eq 11 to 15:

Acoustic impedance $Z_l = \rho\ v_l$ (Eq 11)

Longitudinal modulus $L = \rho\ v_l^2$ (Eq 12)

Shear modulus $S = \rho\ v_t^2$ (Eq 13)

Elastic (Young's) modulus $E = \dfrac{(3L - 4S)S}{(L - S)}$ (Eq 14)

Acoustic reflection coefficient at an interface $R = \left(\dfrac{Z_1 - Z_2}{Z_1 + Z_2}\right)^2$ (Eq 15)

Table 7.3 Acoustic properties of ceramics

Material	Formula	Density (ρ), kg/m³	Sonic velocity, m/s Longitudinal	Shear	Surface	Acoustic impedance (Z_l), Gg/m² · s (a)
Air at 20 °C	$N_2 + O_2$	1.21	340	...	...	4.11×10^{-4}
Alumina	Al_2O_3	3800	10,750	6360	...	40.85
Aluminum	Al	2700	6320	3130	2905	17.06
Aluminum nitride	AlN	3290	10,500	6280	...	34.55
Glass window	SiO_2	2510	5770	3430	...	14.48
Helium at −272 °C	$He_{(l)}$	146	231		...	0.03
Ice	$H_2O_{(s)}$	900	3980	1990	...	3.58
Magnesia	MgO	3842	8850		...	34.00
Silicon nitride	Si_3N_4	3200	10,800	6000	...	34.56
Silicon carbide	α-SiC	3190	12,200	7900	...	38.92
Stainless steel, 302	$Fe_{(\gamma)}$	8030	5660	3120	2780	45.45
Steel, 1020	$Fe_{(\alpha)}$	7710	5890	3240	...	45.41
Superconductor	$YBa_2Cu_3O_{7-x}$	5940	5100	3000	...	30.29
Titanium carbide	TiC	5150	8270	5160	...	42.59
Titanium diboride	TiB_2	4620	12,600	7000	...	58.21
Tungsten carbide	WC	10,000–15,000	6660	3980	...	67–99
Water at 20 °C	$H_2O_{(l)}$	1000	1497	...	...	1.48
Zirconia	ZrO_2	5700	7020	3660	...	40.01

Note: The longitudinal acoustic impedance (Z_l) is equal to the product of longitudinal sonic velocity (v_l) and density (ρ). (a) One gigagram (Gg) = 10^6 kg. Source: Ref 17

where ρ is the density of the ceramic, and v_l and v_t are the longitudinal and transverse sonic velocities, respectively. When air and a solid are the two phases in Eq 15—the acoustic analog to Eq 8—the reflection coefficient is nearly 100%, and the energy transmitted into the specimen ($T = 1 - R$) is negligible. A fluid of higher acoustic impedance than air between the transducer and the solid, such as water, decreases the reflection and greatly increases the transmitted energy.

Scanning acoustic microscopy was used to measure the elastic properties in situ of fine, coated SiC fibers and the SiC matrix in composites formed by chemical vapor infiltration and deposition (Ref 19). The acoustic couplant was a 12.0 mol/L aqueous solution of potash, $KOH_{(aq)}$, which had a lower attenuation than pure water. The couplant density was 1704 kg/m^3 and the sonic velocity was 2244 m/s. The image resolution was approximately 0.7 μm at a sonic frequency of 980 MHz through a lens with an aperture of 50°.

Acoustic microscope manufacturers include Sonoscan Inc. in Illinois, Sonix Inc. in Virginia, and Panametrix Inc. in Massachusetts. A series of books, *Advances in Acoustic Microscopy,* is available from Kluwer Academic/Plenum Publishers in New York. Rochow and Tucker's book has a chapter on acoustic microscopy.

Confocal Laser Scanning Microscopy

A confocal laser scanning microscope (CLSM) uses the monochromatic, in-phase light from a laser in conjunction with a confocal lens system and fluorescence to achieve resolution and depth of field that approach the capabilities of a SEM. The CLSM does not require a vacuum chamber or an electrically conductive specimen, as the SEM does. *Fluorescence* occurs when the light reflected by the specimen is of a longer wavelength than the incident light. In fluorescence microscopy, a black light excites the specimen with invisible ultraviolet incident light to cause fluorescence in the visible spectrum, resulting in better resolution than visible incident light. The confocal lens system improves the depth of field compared to a microscope. The CLSM is primarily used to view surface topography in materials science, and in biology and medicine. A CLSM is shown in Fig. 7.23.

Confocal laser scanning microscopy was used to observe the subsurface microstructures of green ceramic compacts (Ref 20). The green bodies were made transparent by impregnation with methylene chloride (η = 1.74) immersion liquid containing Nile red fluorescent dye. The immersion liquid was chosen so that its refractive index matched that of the ceramic, to minimize reflections from and refraction by the ceramic particles. The specimens were thinned to approximately 200 μm with abrasive paper. The illumination source was an argon ion laser, λ = 488 nm. Two CLSM images of ceramics are shown in Fig. 7.24.

Micrography

A recording of the image is usually the final step in microscopy, regardless of which imaging tool was used to view the specimen. All professional-quality microscopes, electron microscopes, acoustic micro-

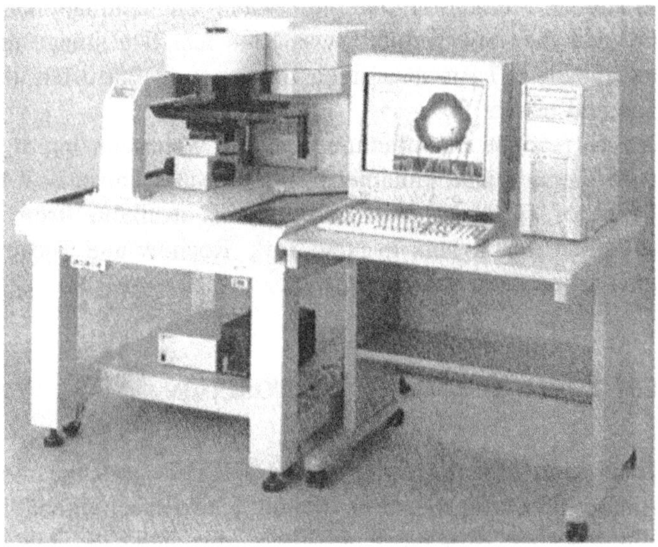

Fig. 7.23 Confocal laser scanning microscope

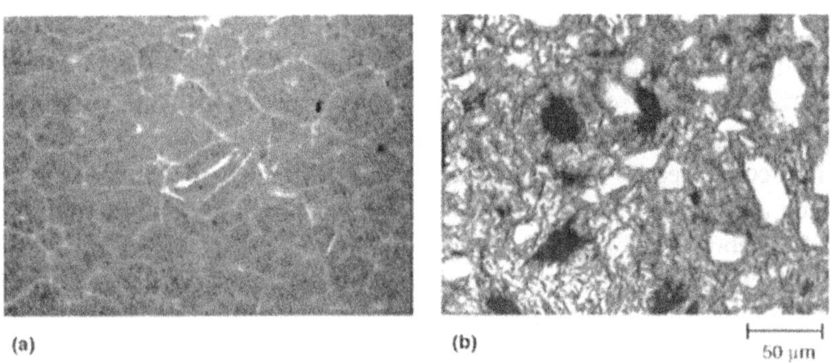

Fig. 7.24 Confocal laser scanning microscopy. (a) CLSM image with fluorescence of a spray-dried alumina compact. The polyvinyl alcohol binder was removed at 1000 °C for 30 min. The larger granules are approximately 50 μm in diameter. (b) CLSM image of hardened gypsum plaster. Courtesy of Y. Saito and K. Uematsu, Nagaoka University of Technology

scopes, and CLSMs have one or more attachments for image recording. The recorded image becomes part of the report of analysis. A photograph of a microscope image is a micrograph.

A Polaroid camera remains the most common type of image-recording device. Polaroid pictures have many advantages, including moderate cost, good resolution, ease of use, worldwide availability, and rapid results. Polaroid film is available in several speeds, in color or black and white, and develops itself on removal from the camera. The whole process, from exposure to a fully developed photograph, typically takes from 1 to 5 min. Some Polaroid film types require a lacquerlike organic coating to preserve the image. Polaroid negatives should be washed in a sodium sulfite solution to remove the gelatinous emulsifier.

Some microscopes have a built-in canister for 35 mm film or an adapter for a 35 mm camera attachment. Standard film records the image by changing the oxidation state of light-sensitive silver halide particles and has all the advantages of Polaroid film except rapid development. Standard photographic film is usually less expensive than Polaroid and yields less wastepaper. Some ceramographic laboratories have their own darkroom, while others simply use the film-development services that are available at most grocery stores and pharmacies. Polaroid and standard film are especially useful for forensic work, where an unretouched negative may be necessary to authenticate an image for legal purposes.

Electronic photography, which records the image in digital form on an integrated circuit chip, has gained popularity in recent years. The digital image can be stored on a disc or in a web page, embedded in a word-processing document or spreadsheet, or edited by a number of software packages. The image can be transmitted almost instantaneously around the world to multiple receivers via e-mail. Digital cameras, scanners, and image analyzers all produce digital electronic images. Digital image formats are discussed in Chapter 11 in conjunction with image analysis. Modern color printers can produce digital images from electronic files that are comparable in resolution, sharpness, and color to the best photographic images.

A micrograph that is included in a report or publication should always have a scale bar instead of or in addition to a magnification. The magnification may change when the photo is reproduced, but the size of the scale bar is changed in proportion to the image shrinkage or enlargement. Most electron microscopes and some other optical instruments can superimpose a scale bar on the photo, as shown in Fig. 7.17, for example. A scale bar can be made, as necessary, with a word processor and a printer. Use the vertical bar and hyphen characters to make a bar approximately 10 mm wide and type the equivalent number of microns next to the scale bar. Print the text, trim it to size with a paper cutter or scissor, and attach it to the image with double-stick tape or glue. A 10 mm bar is convenient, because it is equivalent to 50 μm in a 200× photo, 20 μm at 500×, 10 μm at 1000×, and so on. For a photo taken from a macrostand where the magnification is 1× or less, as is commonly done in failure analysis, set a ruler next to the object so that the scale of the ruler shows up in the image. Use an image-editing program, for example, Adobe Photoshop or Microsoft Paint, to superimpose a scale bar on an electronic image.

Some of the images in this text were recorded directly as digital images. Most were originally Polaroid micrographs that were scanned and computer edited.

REFERENCES

1. G.F. Vander Voort, Optical Microscopy, *Metallography and Microstructures,* Vol 9, *ASM Handbook,* American Society for Metals, 1985, p 75–76, 80, 81
2. W.C. McCrone, L.B. McCrone and J.G. Delly, *Polarized Light Microscopy,* McCrone Research Institute, 1984, p 33, 135
3. G. Petzow and G. Elssner, Etching, *Metallography and Microstructures,* Vol 9, *Metals Handbook* 9th ed., American Society for Metals, 1985, p 58
4. T.G. Rochow and P.A. Tucker, *Introduction to Microscopy by Means of Light, Electrons, X-Rays, or Acoustics,* 2nd Ed., Plenum Press, 1994, p 43–49, 339–344
5. P. Perlman, *Basic Microscope Techniques,* Chemical Publishing Co., Inc., 1971, p 168
6. T.K. Bierlein, H.W. Newkirk, Jr., and B. Mastel, Etching of Refractories and Cermets by Ion Bombardment, *J. Am. Ceram. Soc.,* Vol 41 (No. 11), June 1958, p 196–200
7. A.R. Marder, Replication Microscopy Techniques for NDE, *Nondestructive Evaluation and Quality Control,* Vol 17, *Metals Handbook* (9 ed.), ASM International, 1989, p 52–56
8. H.E. Exner, Scanning Electron Microscopy, *Metallography and Microstructures,* Vol 9, *ASM Handbook,* American Society for Metals, 1985, p 89–102
9. J.I. Goldstein, D.E. Newbury, P. Echlin, D.C. Joy, C. Fiori, and E. Lifshin, *Scanning Electron Microscopy and X-Ray Microanalysis,* Plenum Press, 1981, p 17
10. J.D. Verhoeven, Scanning Electron Microscopy, *Materials Characterization,* Vol 10, *ASM Handbook,* American Society for Metals, 1986, p 490–515
11. R.E. Chinn, "Transient Glass-Phase Processing of $SrAl_2Si_2O_8$-Alumina Ceramics," Thesis T-4685, Colorado School of Mines, 1995, p 113
12. A.H. Hunt and R.E. Chinn, A Ceramographic Evaluation of Chromia Refractories Corroded by Slag, *Structure,* Vol 37, 2001, p 6–10

13. K.F.J. Heinrich and D.E. Newbury, Electron Probe X-Ray Microanalysis, *Materials Characterization,* Vol 10, *ASM Handbook,* American Society for Metals, 1986, p 516–535
14. M.Rühle, Transmission Electron Microscopy, *Metallography and Microstructures,* Vol 9, *Metals Handbook* (9th ed.), American Society for Metals, 1985, p 103–122
15. A.D. Romig, Jr., Analytical Transmission Electron Microscopy, *Materials Characterization,* Vol 10, *ASM Handbook,* American Society for Metals, 1986, p 429–489
16. L.W. Kessler, Acoustic Microscopy: Industrial and Electronics Applications, *Ultrasonic Testing, Nondestructive Testing Handbook,* 2nd ed., American Society for Nondestructive Testing, Inc., 1991, p 797–817
17. A.S. Birks, R.E. Green, Jr., and P. McIntire, Ed., *Ultrasonic Testing, Vol 7, Nondestructive Testing Handbook,* 2nd ed., American Society for Nondestructive Testing, Inc., 1991, p 840–849
18. A. Vary, Microstructure and Diffuse Discontinuities, *Ultrasonic Testing, Nondestructive Testing Handbook,* 2nd ed., American Society for Nondestructive Testing, Inc., 1991, p 403–409
19. B. Cros, P. Combette, and J. Attal, Elastic Property Measurements on CVI SiC/SiC Minicomposites by Acoustic Microscopy, *J. Am. Ceram. Soc.,* Vol 83 (No. 4), 2000, p 809–816
20. Y. Saito, S. Tanaka, N. Uchida, and K. Uematsu, CLSM for Ceramic Green Microstructure, *Am. Ceram. Soc. Bull.,* Vol 81 (No. 1), 2002, p 35–38

SUGGESTED READING

- S.H. Cohen and M.L. Lightbody, Ed., *Atomic Force Microscopy/Scanning Tunneling Microscopy 3,* Kluwer Academic Publishers, 1999
- L. Dillinger, "Stereo-Pairs from Low Power Microscopes," Met-Tips 8, Leco Corp., 1985
- L. Dillinger, "In-Situ Metallography," Met-Tips 3, Leco Corp., 1989
- J.I. Goldstein, D.E. Newbury, P. Echlin, D.C. Joy, A.D. Romig, Jr., C.E. Lyman, C. Fiori, and E. Lifshin, *Scanning Electron Microscopy and X-Ray Microanalysis,* 2nd ed., Kluwer Academic/Plenum Publishers, 1996
- H. Insley and V.D. Fréchette, *Microscopy of Ceramics and Cements,* Academic Press Inc., 1955
- W.C. McCrone, L.B. McCrone, and J.G. Delly, *Polarized Light Microscopy,* McCrone Research Institute, 1984
- L.E. Murr, *Electron Optical Applications in Materials Science,* McGraw-Hill Book Co., 1970

- T.G. Rochow and P.A. Tucker, *Introduction to Microscopy by Means of Light, Electrons, X-Rays, or Acoustics,* 2nd ed., Plenum Press, 1994
- G.E. Vander Voort, *Metallography: Principles and Practice,* American Society for Metals, 1984
- M. Von Heimendahl, *Electron Microscopy of Materials: An Introduction,* U.E. Wolff, Trans., Academic Press, Inc., 1980
- D.B. Williams and C.B. Carter, *Transmission Electron Microscopy,* Plenum Press, 1996

CHAPTER 8

Atlas of Ceramic Microstructures

MICROSTRUCTURES of selected ceramics from various manufacturers are shown in Fig. 8.1 to 8.91. The mean grain sizes ("GS" in the captions) were determined by the circle-intercept method of ASTM E 112, with a correction factor A_1 = 1.0, unless otherwise noted.

Alumina

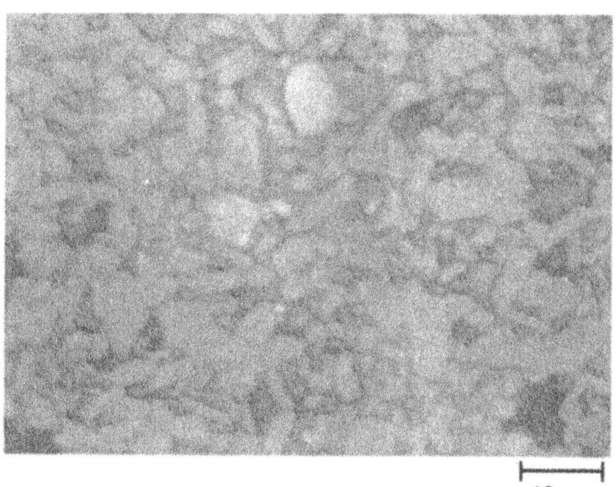

Fig. 8.1 Dry-pressed 85% Al_2O_3 made from reclaimed scrap and thermally etched 5 min at 1500 °C. Mean GS = 2.2 μm

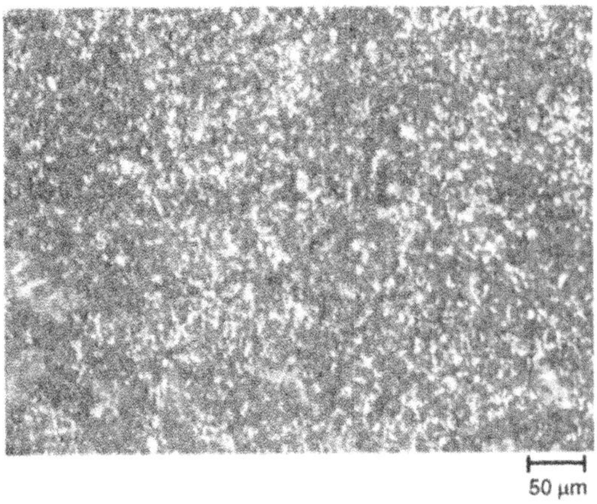

Fig. 8.2 85% Al_2O_3 thin section

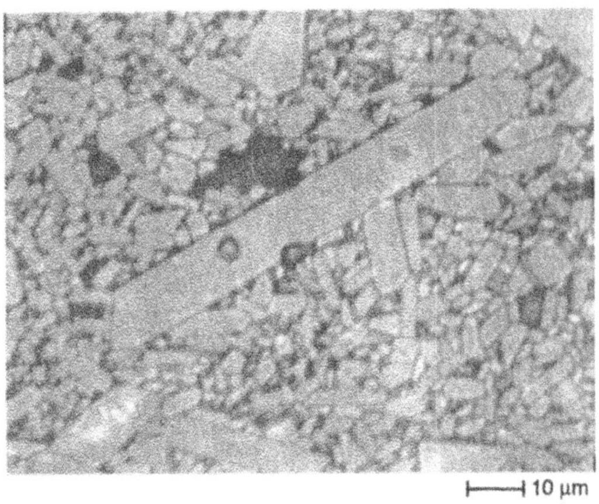

Fig. 8.3 Dry-pressed 90% Al_2O_3 tinted brown with MnO_2 and thermally etched 5 min at 1500 °C. Mean GS = 3.3 µm

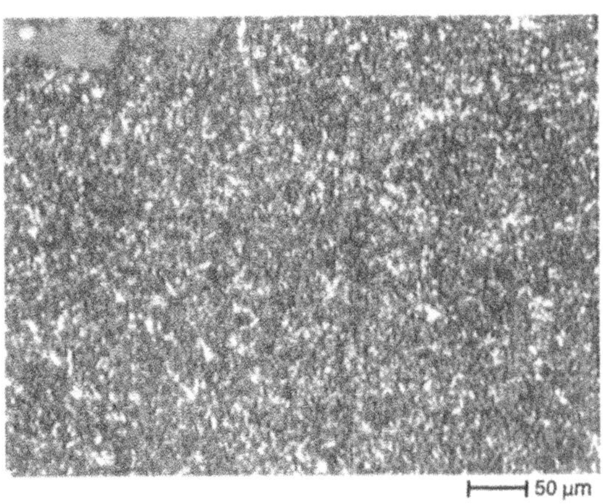

Fig. 8.4 90% Al_2O_3 thin section

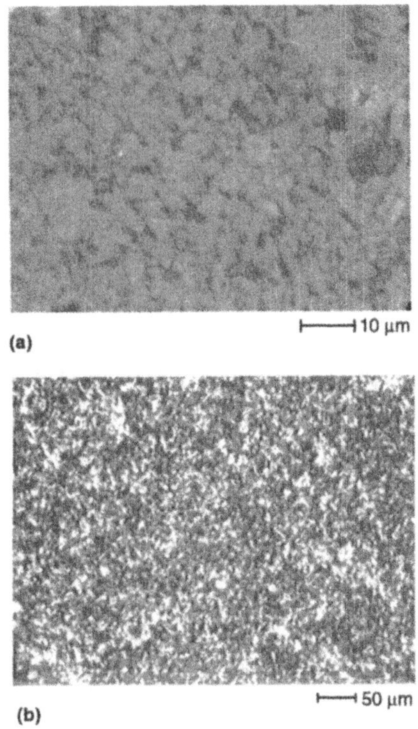

Fig. 8.5 (a) Slip-cast 90% Al_2O_3 thermally etched 5 min at 1500 °C. Mean GS =2.7 µm. (b) 90% Al_2O_3 thin section from the same process as (a)

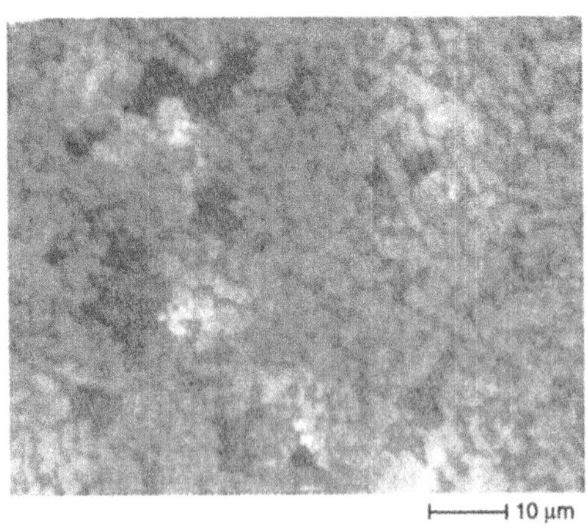

Fig. 8.6 Italian 90% Al_2O_3 grinding ball thermally etched 5 min at 1500 °C. Mean GS = 2.4 µm

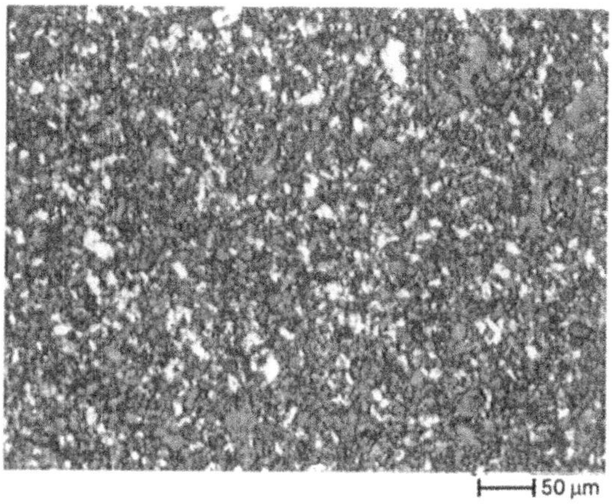

Fig. 8.7 Thin section of a 94% Al_2O_3 valve compnent

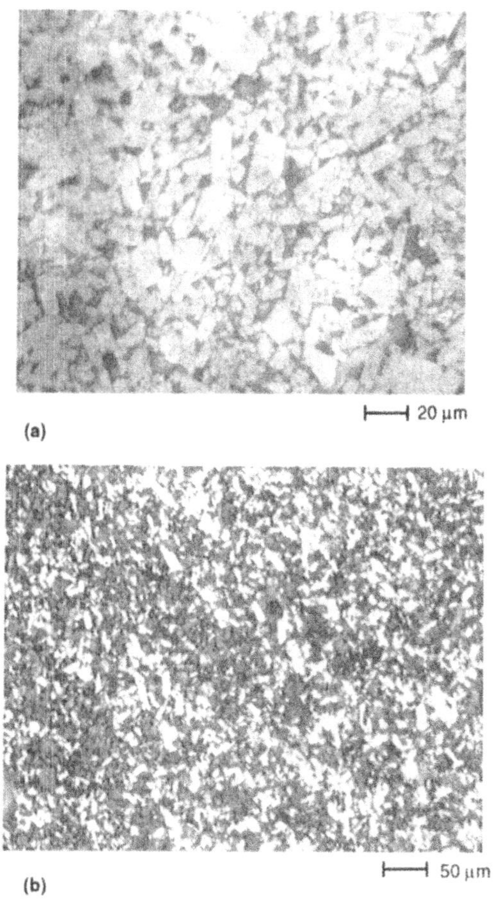

Fig. 8.8 (a) Dry-pressed 94% Al_2O_3, thermally etched. Mean GS = 7.0 µm. Min GS < 2 µm. Max GS ≈ 50 µm × 65 µm. (b) 94% Al_2O_3 thin section from the same process as (a)

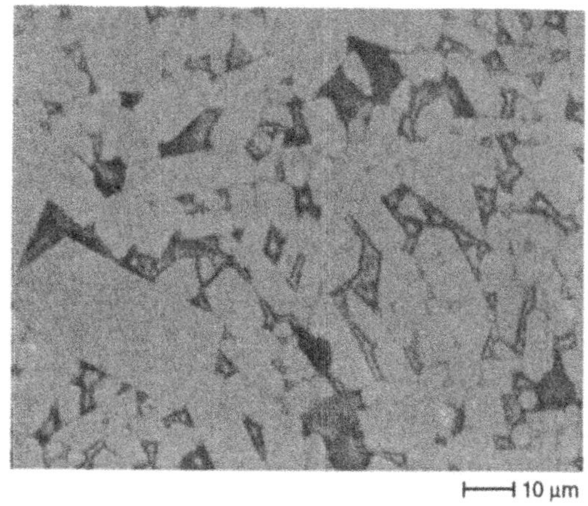

Fig. 8.9 Dry-pressed 94% Al_2O_3 thermally etched 5 min at 1500 °C. Mean GS = 5.7 µm

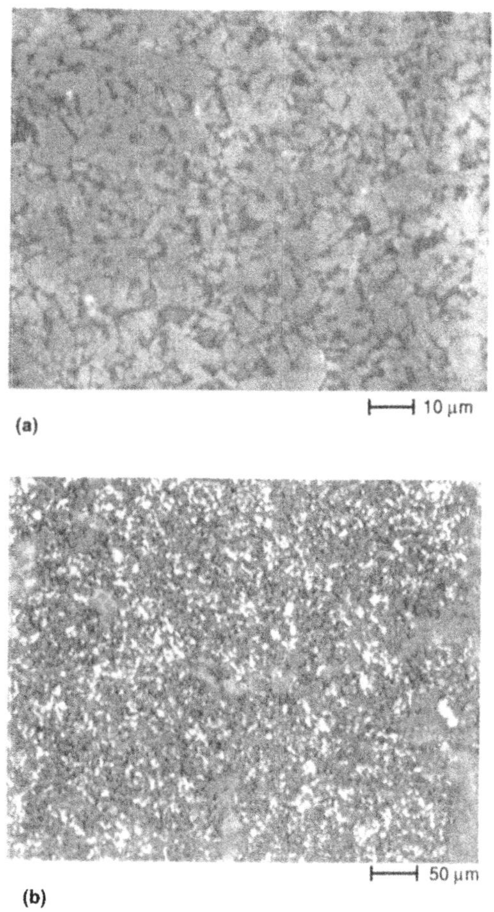

Fig. 8.10 (a) 95% Al_2O_3 thermally etched 5 min at 1500 °C. (b) 95% Al_2O_3 thin section from the same process as (a)

Fig. 8.11 96% Al_2O_3 thermally etched 10 min at 1500 °C and gold sputter coated. The mean grain size is 3.1 µm by the circle-intercept method, 3.3 µm by the lineal intersection method, and 3.1 µm by image analysis (Ref 1).

Fig. 8.12 Isopressed "opaque" (doped with MnO_2) 96% Al_2O_3 pump shaft from Scotland that fractured during sintering, thermally etched 3 min at 1500 °C

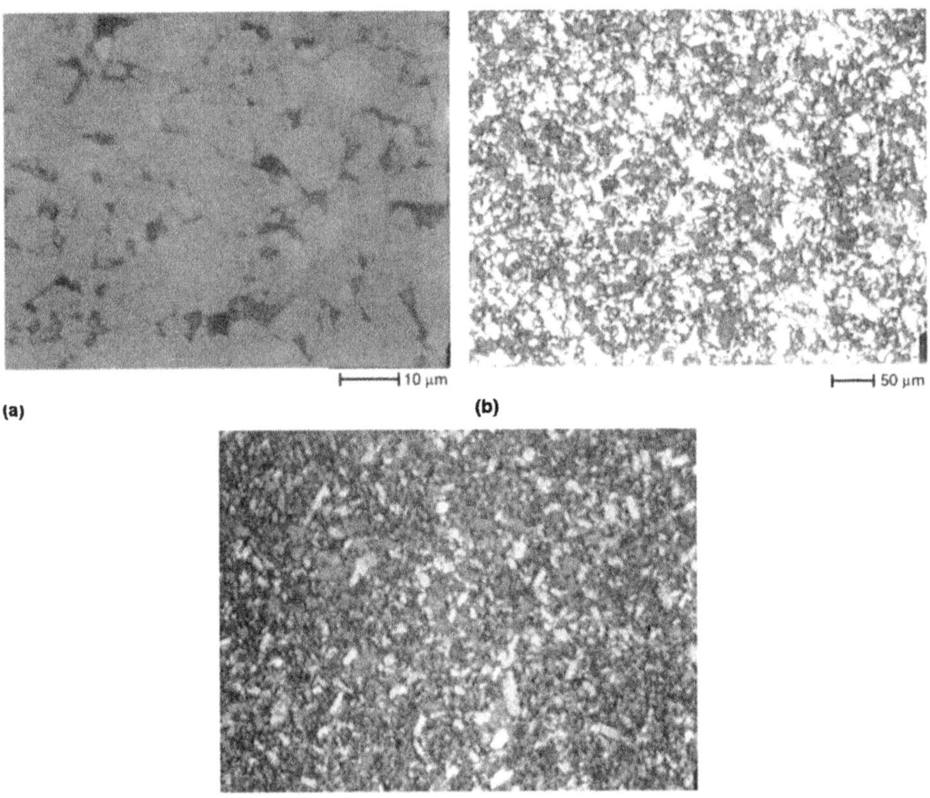

(a) (b) (c)

Fig. 8.13 (a) Dry-pressed 96% Al_2O_3 thermally etched 5 min at 1500 °C. Mean GS = 4.3 µm. (b) 96% Al_2O_3 thin section from the same process as (a). (c) 96% Al_2O_3 thin section from the same process as (a) and (b). Polaroid color photo viewed through a full-wavelength compensator. For color version of Fig. 8.13(c), see endsheets of book.

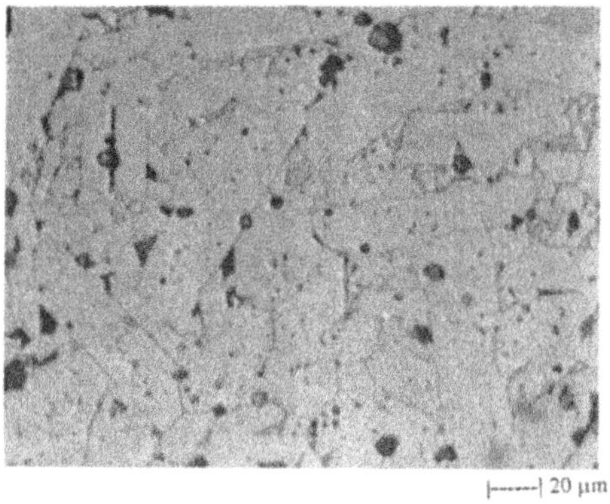

Fig. 8.14 Experimental 96% Al_2O_3 with intergranular monoclinic $SrAl_2Si_2O_8$ (Ref 2). Thermally etched 5 min at 1500 °C and sputter coated. Mean GS = 8.5 µm by image analysis. The ceramic powder was prepared by the method described in the section "Laboratory Fabrication of Ceramics" in Chapter 2.

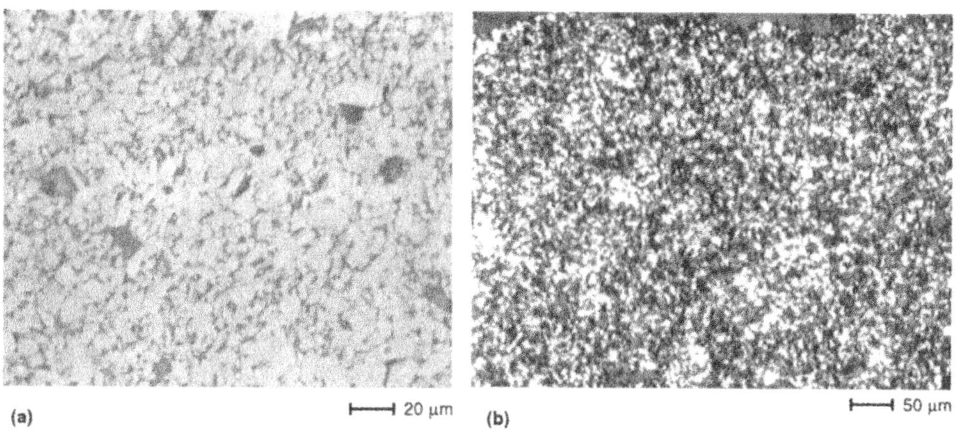

Fig. 8.15 (a) Dry-pressed 96% Al_2O_3 from a liquid-phase sintering process, thermally etched 5 min at 1500 °C and gold sputter coated. (b) 96% Al_2O_3 thin section from the same process as (a)

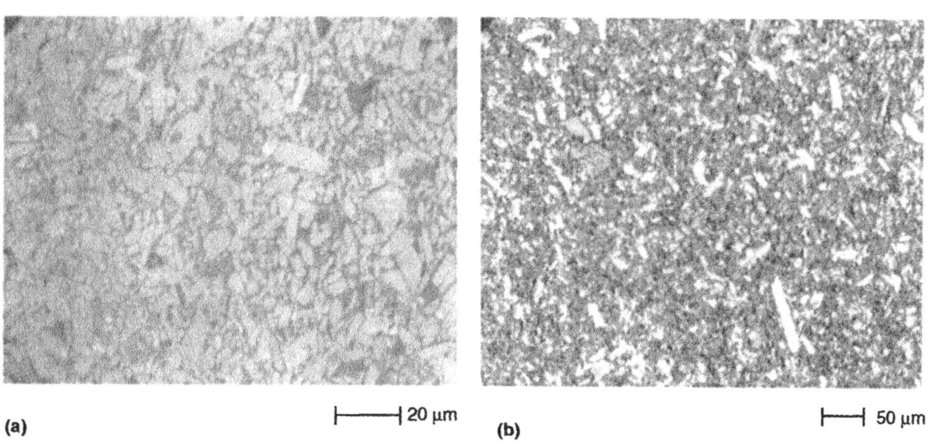

Fig. 8.16 (a) Dry-pressed 96% Al_2O_3 thermally etched 5 min at 1500 °C. Mean GS = 3.1 µm. (b) 96% Al_2O_3 thin section from the same process as (a)

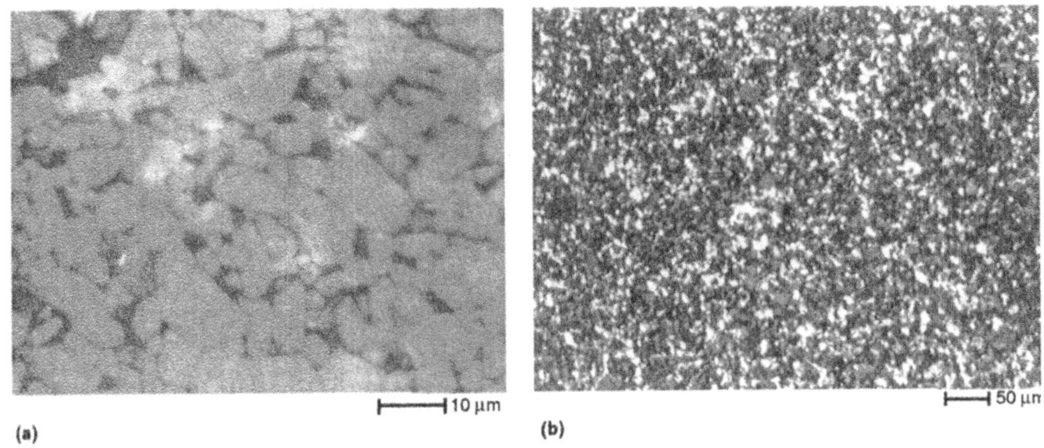

Fig. 8.17 (a) Isopressed 96% Al_2O_3 thermally etched 5 min at 1500 °C. Mean GS = 4.1 µm. (b) 96% Al_2O_3 thin section from the same process as (a)

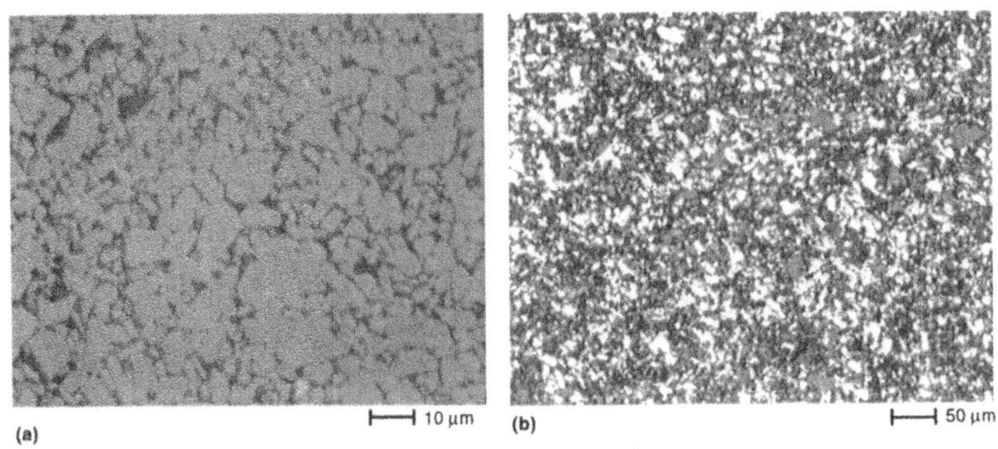

Fig. 8.18 (a) Isopressed 96% Al_2O_3 thermally etched 5 min at 1500 °C. Mean GS = 3.1 µm. (b) 96% Al_2O_3 thin section from the same process as (a)

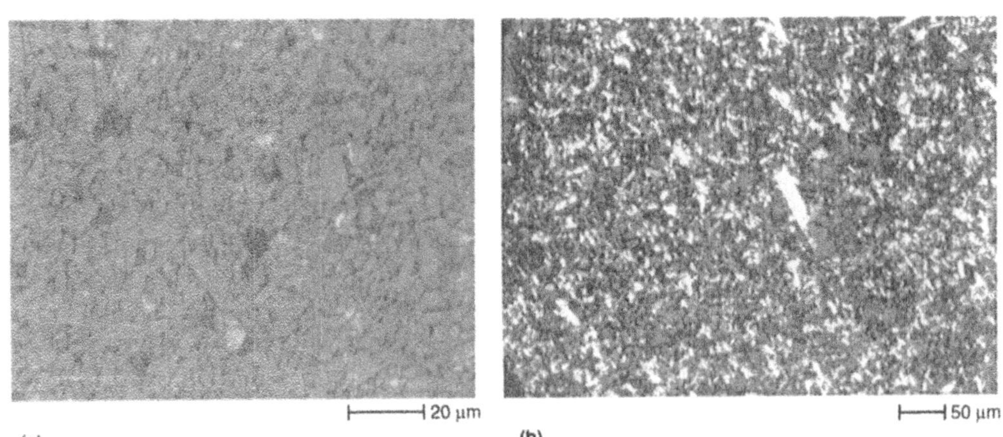

Fig. 8.19 (a) Isopressed "opaque" (doped with MnO_2) 96% Al_2O_3 thermally etched 5 min at 1500 °C. Mean GS = 3.0 µm. (b) 96% Al_2O_3 thin section from the same process as (a)

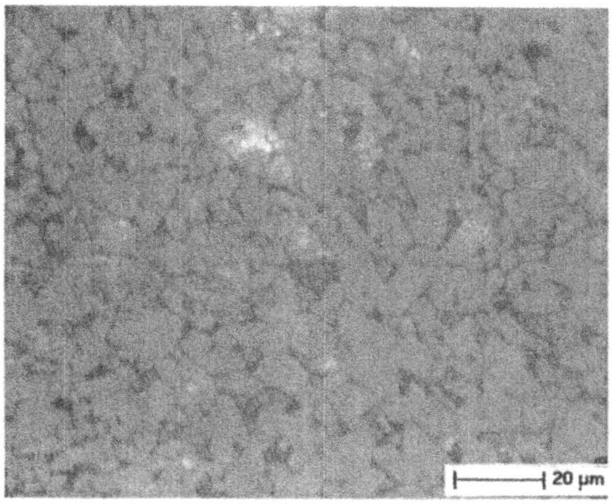

Fig. 8.20 Roll-compacted 96% Al_2O_3 substrate made in the United States from essentially the same process as Fig. 8.22(a) and thermally etched 5 min at 1500 °C. Mean GS = 3.5 μm

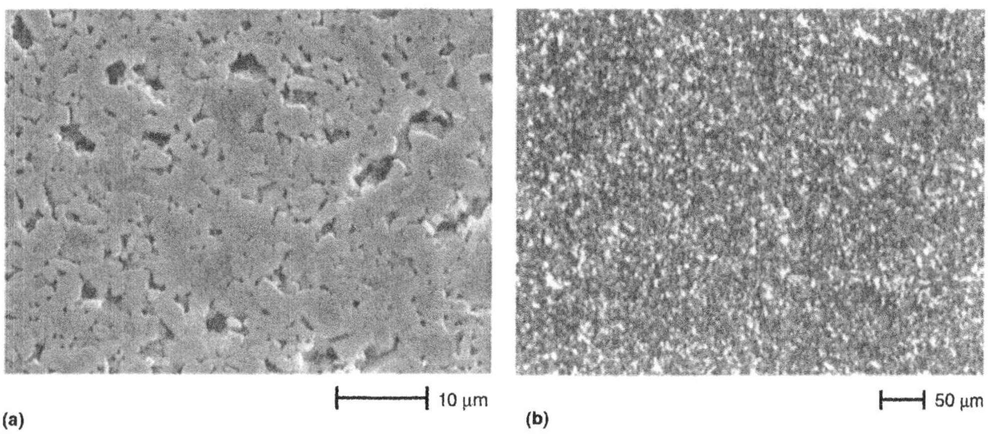

Fig. 8.21 (a) Scanning electron microscope photo of a dry-pressed 96% Al_2O_3 chemically etched 30 s in boiling phosphoric acid. Mean GS = 2.8 μm. The grains are too small for a good microscope image. (b) 96% Al_2O_3 thin section from the same process as (a)

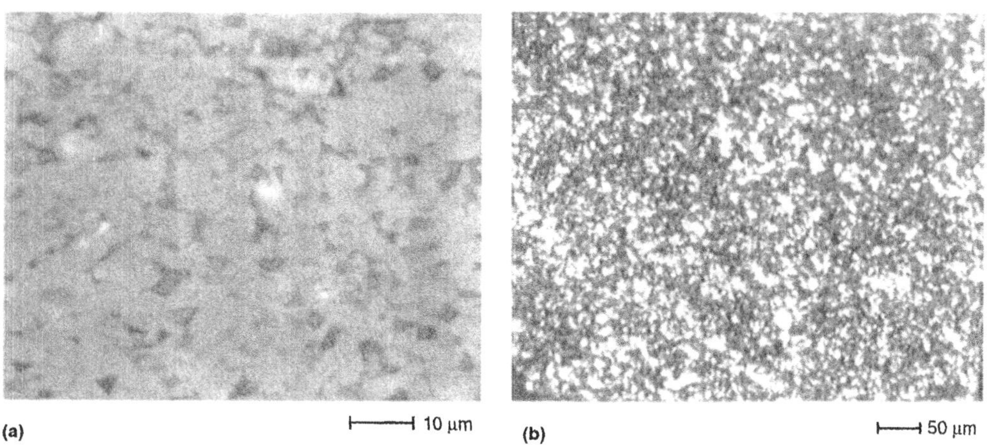

Fig. 8.22 (a) 96% Al_2O_3 roll-compacted substrate from Brazil thermally etched 5 min at 1500 °C. Mean GS = 4.0 μm. (b) 96% Al_2O_3 thin section from the same process as (a)

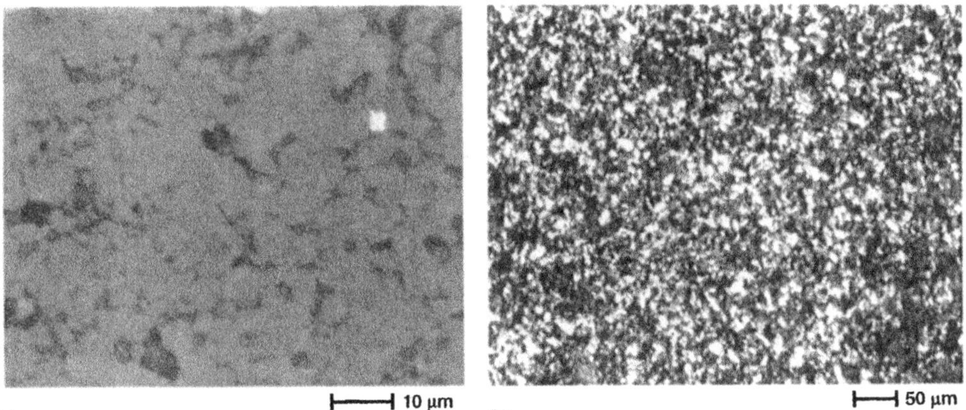

(a) (b)

Fig. 8.23 (a) Roll-compacted 96% Al_2O_3 substrate thermally etched 3 min at 1500 °C. Mean GS = 3.7 µm. (b) 96% Al_2O_3 thin section from the same process as (a)

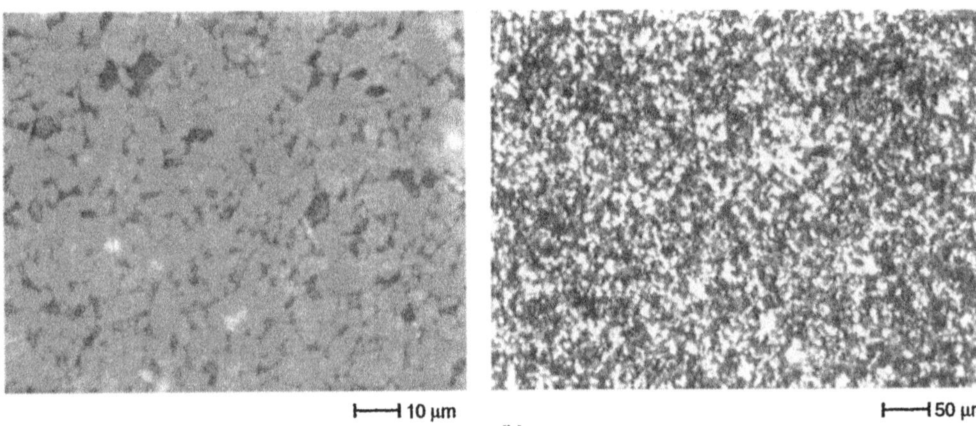

(a) (b)

Fig. 8.24 (a) Roll-compacted 96% Al_2O_3 substrate thermally etched 5 min at 1500 °C. Mean GS = 4.1 µm. (b) 96% Al_2O_3 thin section from the same process as (a)

Fig. 8.25 Transmission electron microscope photo of an experimental 96% Al_2O_3 thin section. The secondary phases are plagioclase and enstatite ($MgSiO_3$). Courtesy of C.P. Doğan

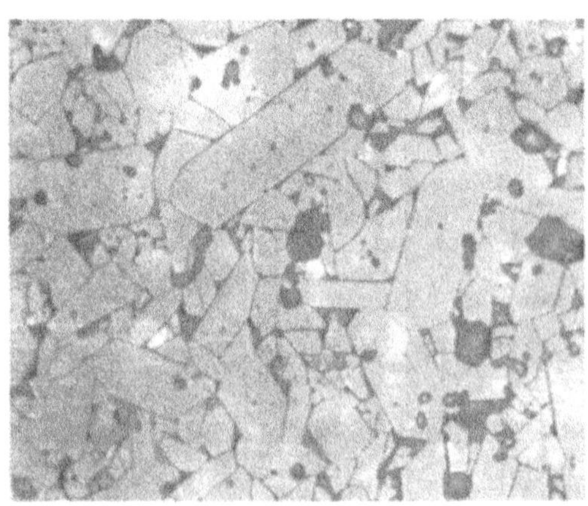

Fig. 8.26 97% Al_2O_3 chemically etched 30 s in boiling phosphoric acid. Mean GS = 12 µm

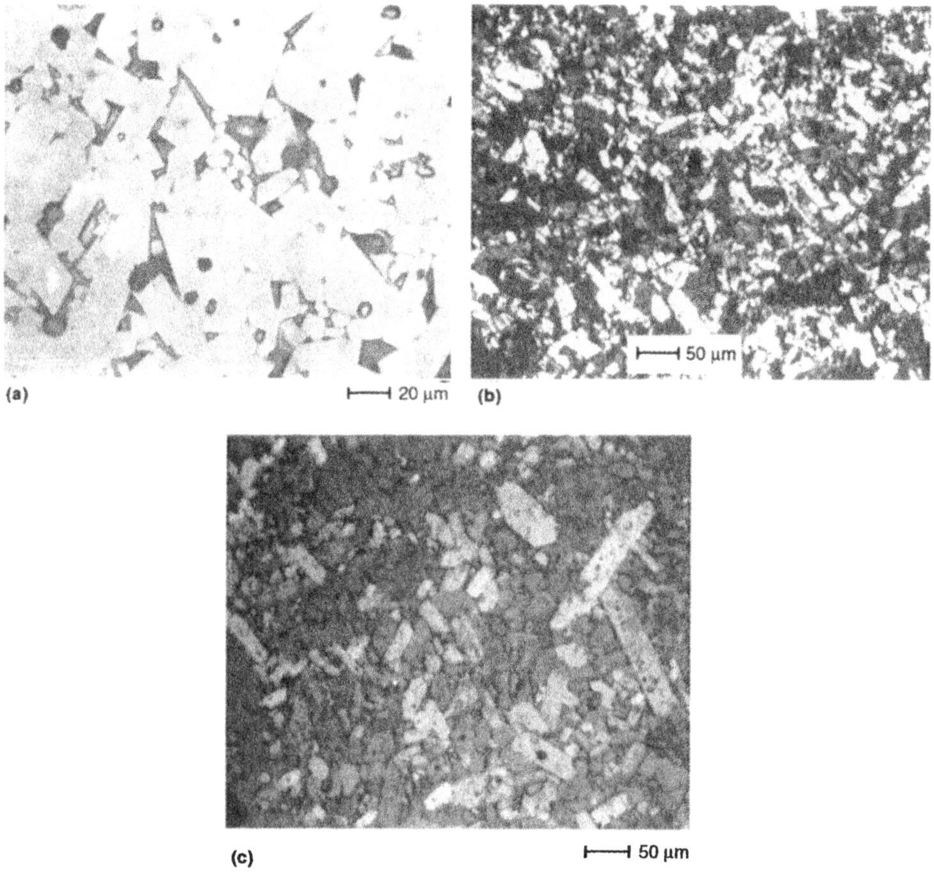

Fig. 8.27 (a) Isopressed 98% Al_2O_3 thermally etched 5 min at 1500 °C. Mean GS = 14 μm. (b) 98% Al_2O_3 thin section from the same process as (a). (c) 98% Al_2O_3 thin section from the same process as (a) and (b). Polaroid color photo viewed through a full-wavelength compensator. For color version of Fig. 8.27(c), see endsheets of book.

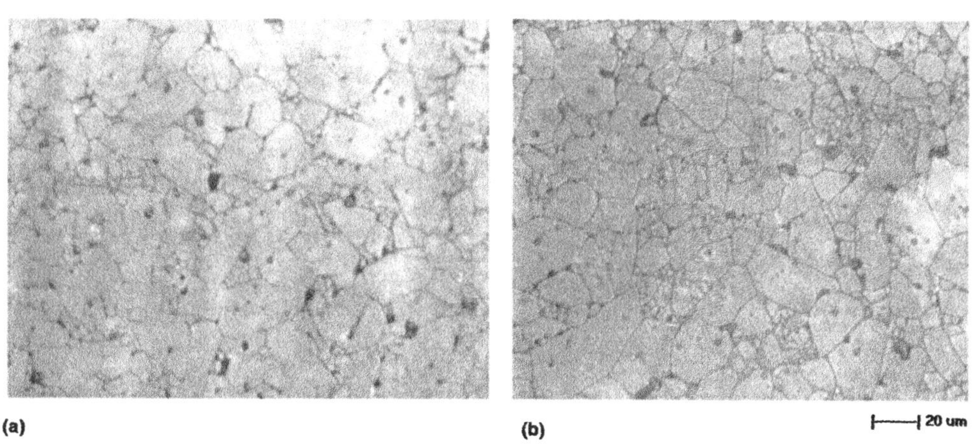

Fig. 8.28 (a) 99.5% Al_2O_3, thermally etched. (b) Iso-pressed 99.5% Al_2O_3, thermally etched 15 min at 1500 °C. Mean grain size = 7.7 μm

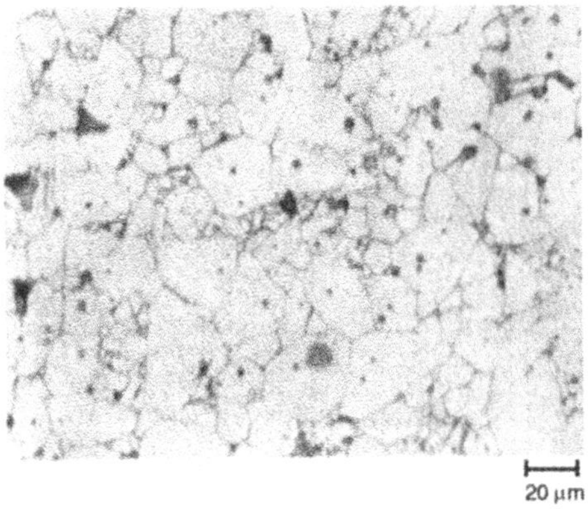

Fig. 8.29 99.5% Al_2O_3 thermally etched 30 min at 1500 °C and gold sputter coated. The mean grain size is 7.7 µm by the circle-intercept method, 8.3 µm by the lineal intersection method, and 7.2 µm by image analysis. Source: Ref 1

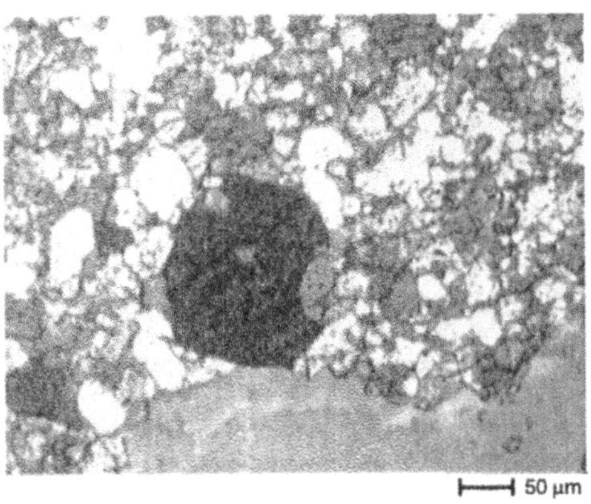

Fig. 8.30 Overfired 99.5% Al_2O_3 mandrel thin section. The stage was rotated to the extinction angle of the oversize grain in the center. Microcracks are faintly visible in the neighboring grains.

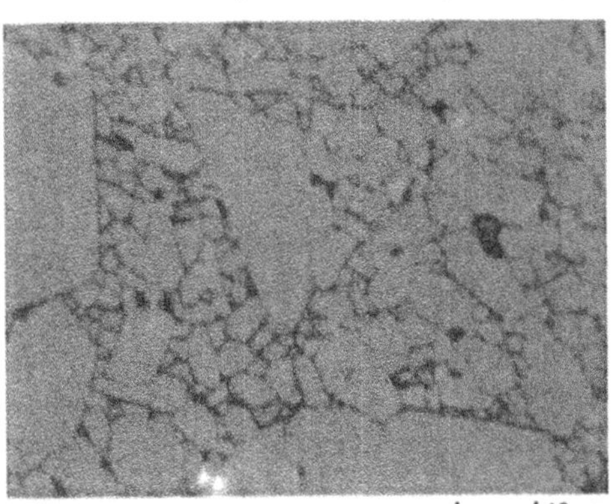

Fig. 8.31 Isopressed 99.5% Al_2O_3 thermally etched 13 min at 1500 °C. The large-trapezoidal grain near the center and its surrounding smaller grains demonstrate a "necklaced" microstructure.

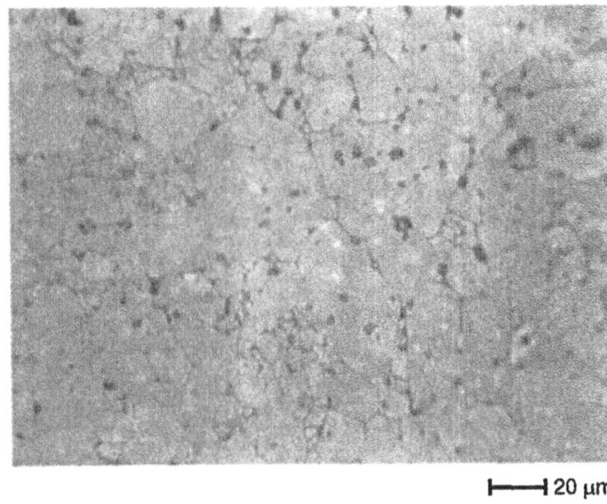

Fig. 8.32 Japanese 99.5% Al_2O_3 piston sleeve thermally etched. Mean GS = 6.6 µm

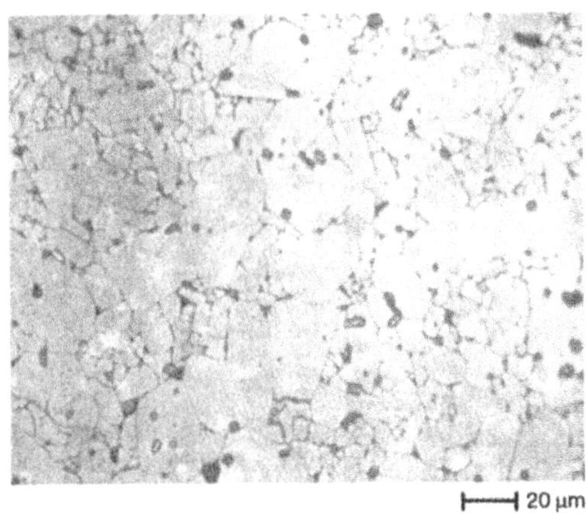

Fig. 8.33 99.5% Al_2O_3 thermally etched 15 min at 1500 °C in flowing N_2. Mean GS = 6.4 µm

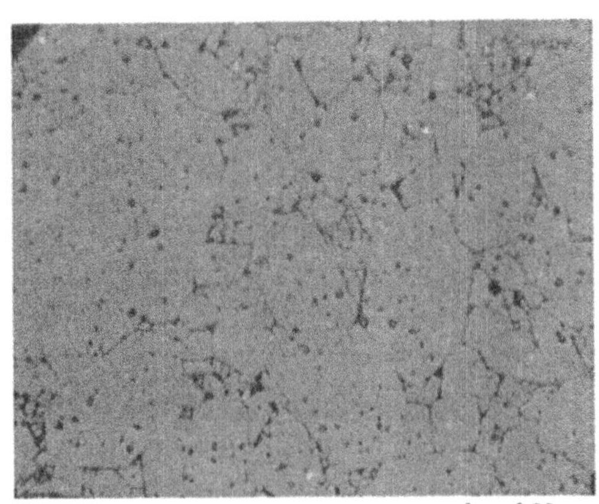

Fig. 8.34 99.5% Al_2O_3 thermally etched 13 min at 1500 °C and photographed through a yellow-green ($\lambda \approx 546$ nm) filter

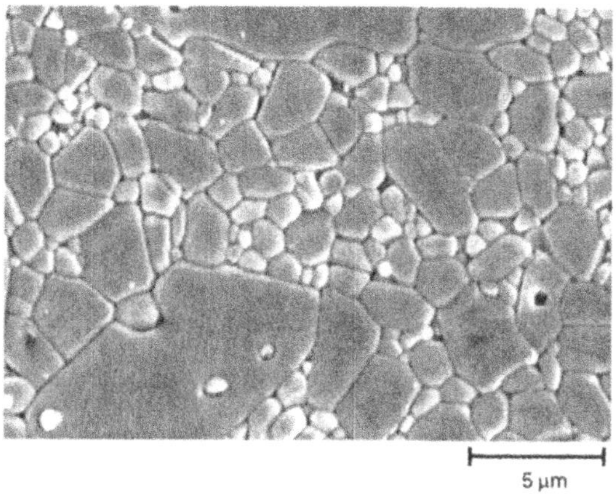

Fig. 8.35 Very fine-grain 99.5% Al_2O_3 tube, thermally etched and photographed by SEM

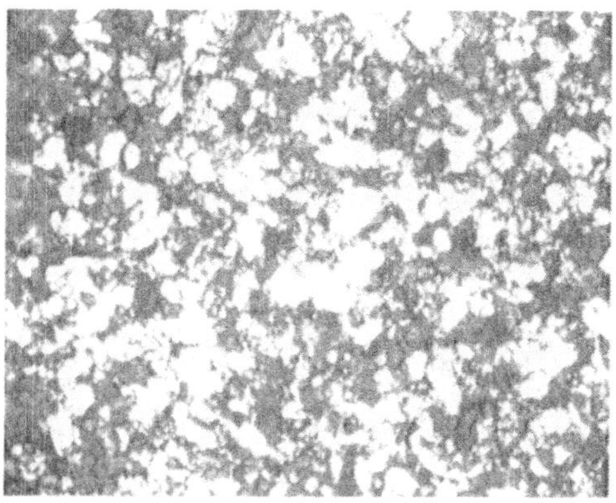

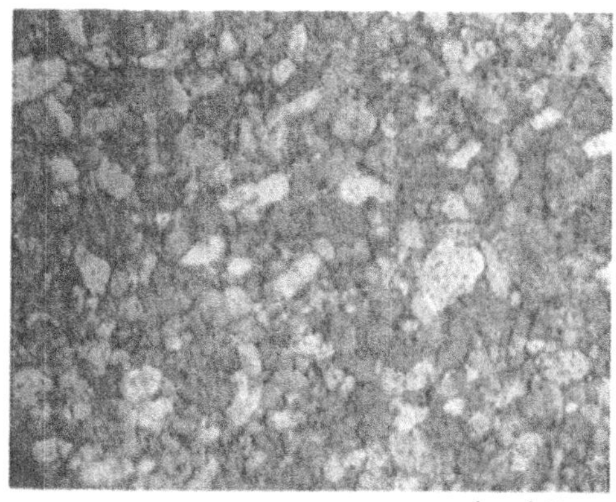

Fig. 8.36 (a) Isopressed 99.5% Al_2O_3 thin section. (b) 99.5% Al_2O_3 thin section from the same process as (a). Polaroid color photo viewed through a full-wavelength compensator. For a color version of Fig. 8.36(b), see end sheets of book.

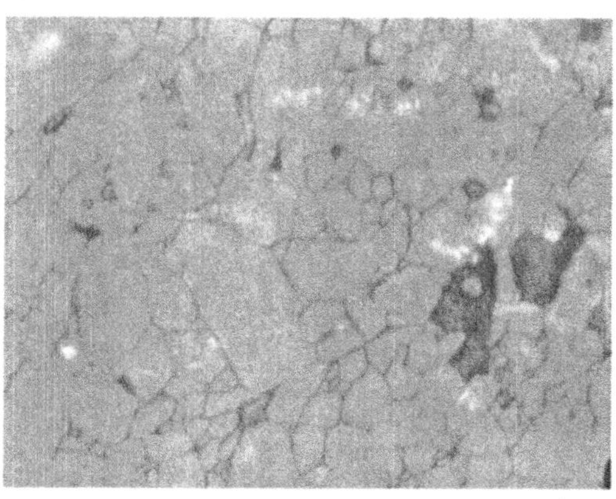

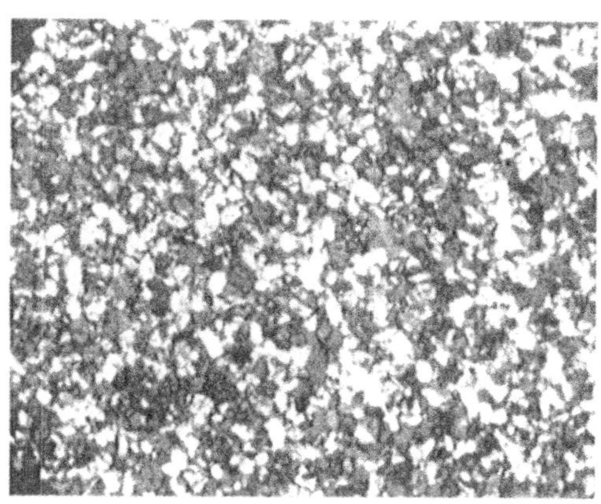

Fig. 8.37 (a) 99.7% Al_2O_3 thermally etched 15 min at 1500 °C. Mean GS = 5.5 µm. (b) 99.7% Al_2O_3 thin section from the same process as (a)

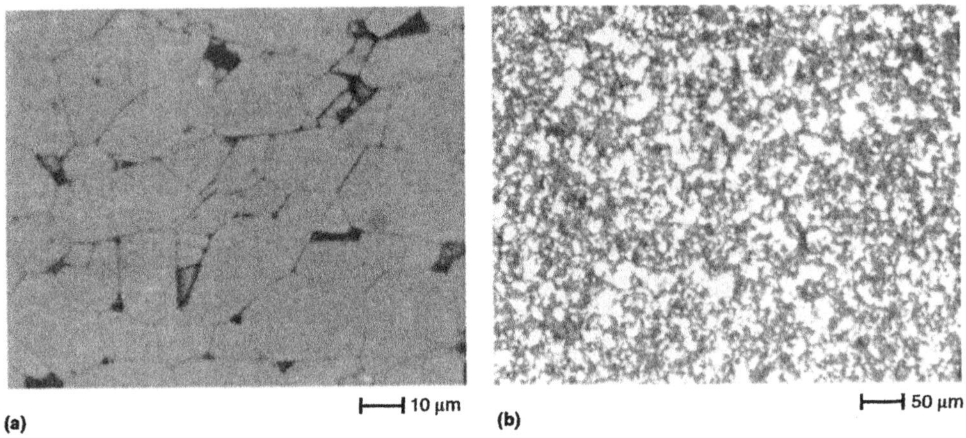

Fig. 8.38 (a) 99.8% Al_2O_3 thermally etched 10 min at 1500 °C and viewed in differential interference contrast. Mean GS = 11 μm. (b) 99.8% Al_2O_3 thin section from the same process as (a)

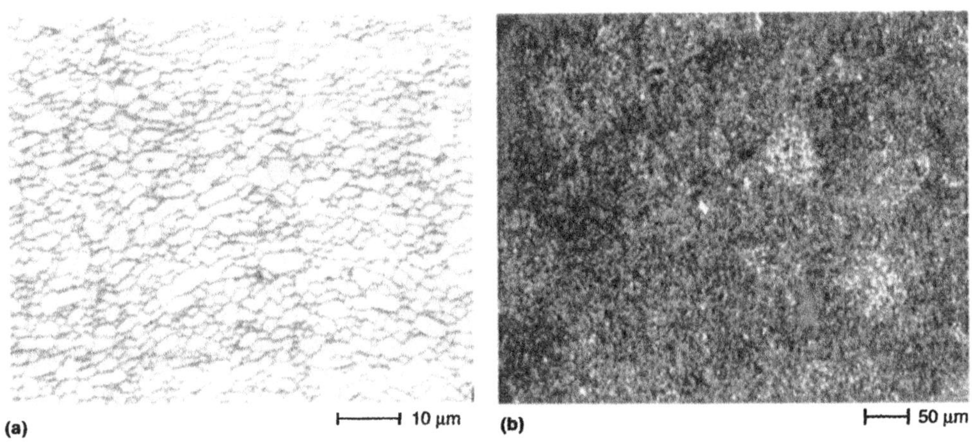

Fig. 8.39 (a) Extruded 99.9% Al_2O_3 tube thermally etched 12 min at 1500 °C. Mean GS = 1.7 μm. (b) 99.9% Al_2O_3 thin section from the same process as (a).

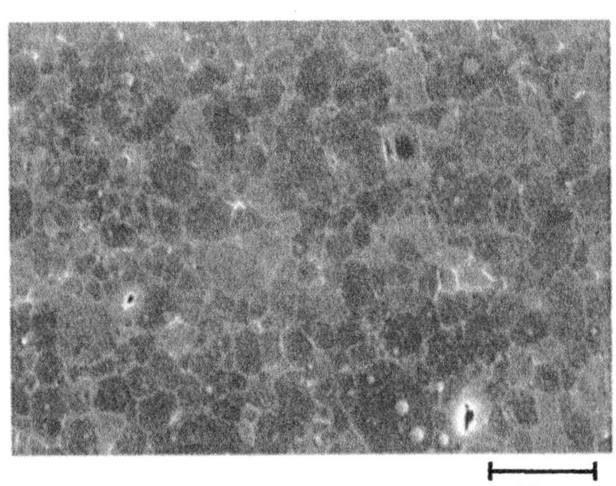

Fig. 8.40 99.9% Al_2O_3 sintered at 1600 °C in a reducing atmosphere. Thermally etched 15 min at 1500 °C. SEM photo. Mean GS = 2.1 μm

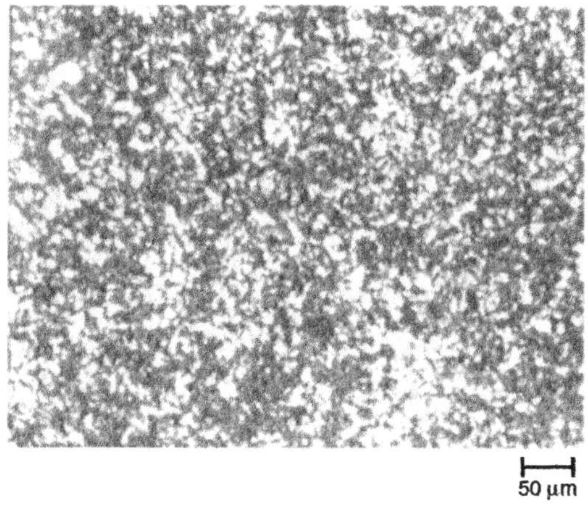

Fig. 8.41 99.9% Al_2O_3 thin section

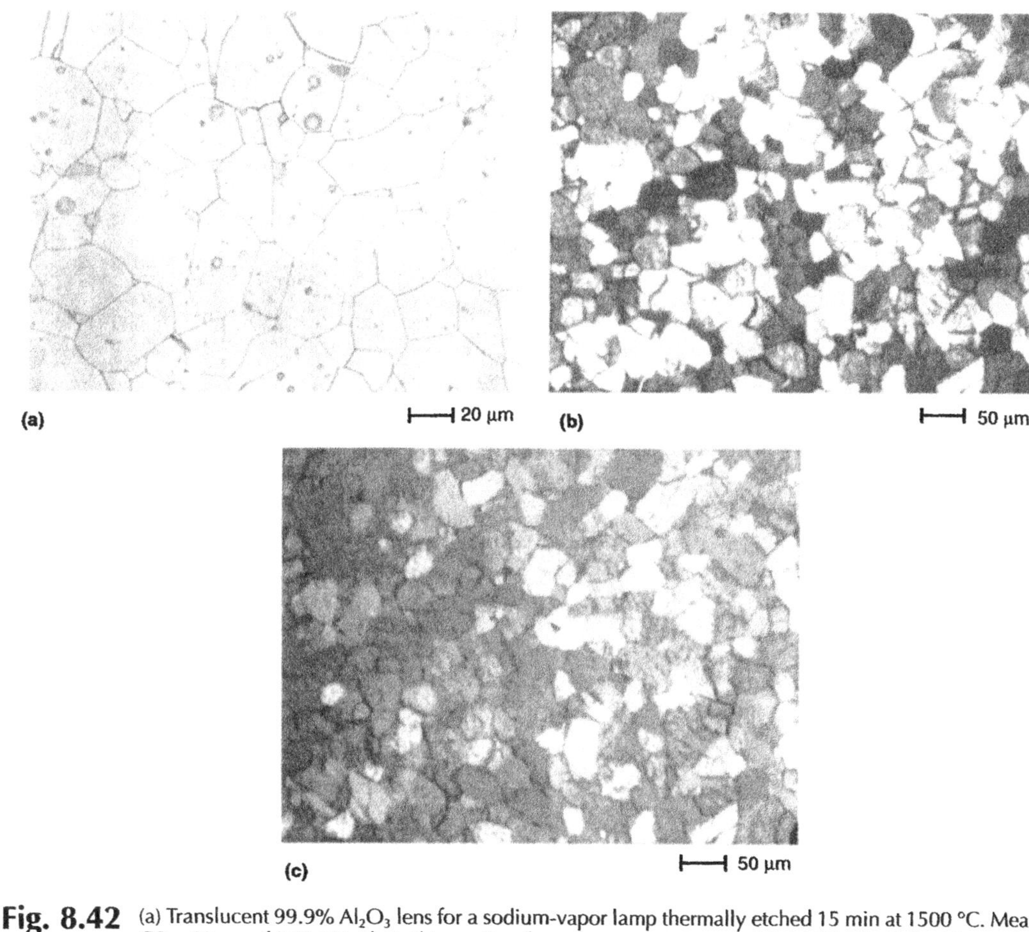

Fig. 8.42 (a) Translucent 99.9% Al_2O_3 lens for a sodium-vapor lamp thermally etched 15 min at 1500 °C. Mean GS = 19 µm. (b) 99.9% Al_2O_3 thin section from the same process as (a). (c) 99.9% Al_2O_3 thin section from the same specimen as (a) and (b). Polaroid color photo viewed through a full-wavelength compensator. For color version of Fig. 8.42(c), see end sheets of book.

Borides

Fig. 8.43 Titanium boride chemically etched 60 s in 1.0 N sulfuric acid. Viewed in differential interference contrast

Carbides

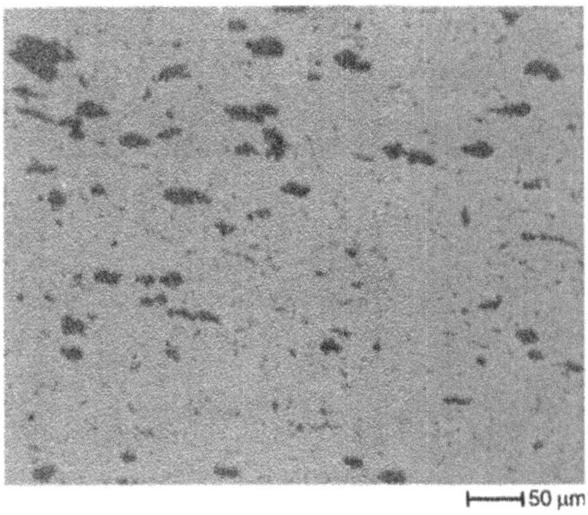

Fig. 8.44 Boron carbide not etched

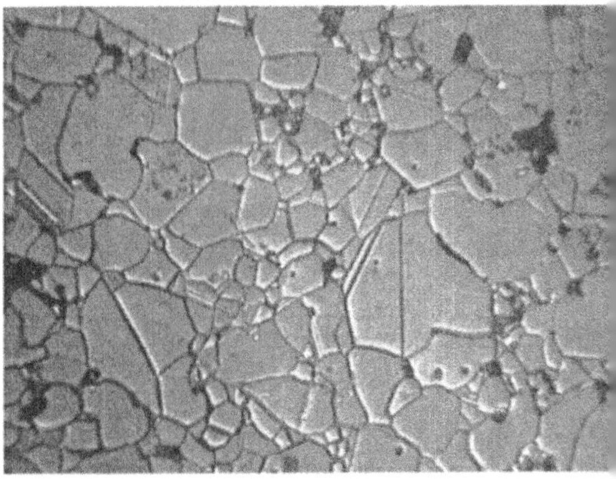

Fig. 8.45 Boron carbide electrolytically etched 6 min in 1% $KOH_{(aq)}$. Stainless steel electrodes, emf = 5 V_{DC}, I = 0.5 A_{DC}, current density = 1.2 A/cm^2, room temperature

Fig. 8.46 Titanium carbide heat tinted 5 min in a Bunsen flame. Murakami's etchant had no effect.

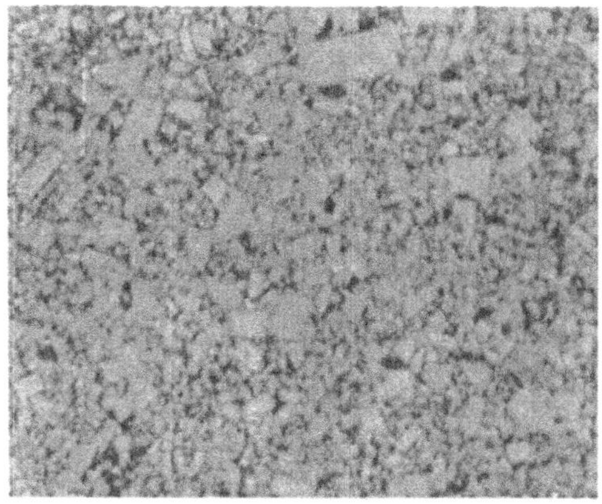

Fig. 8.48 Cobalt-bonded tungsten carbide (WC) die from a beverage can bodymaker chemically etched 10 s in Murakami's etchant at room temperature. Viewed in oil immersion

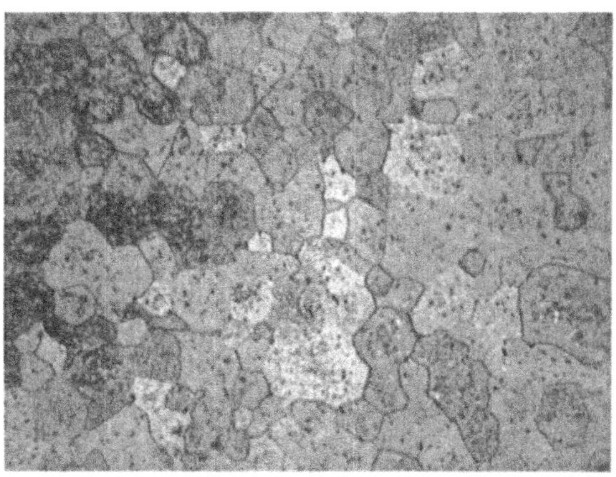

Fig. 8.47 Titanium carbide electrolytically etched 4 min in 1% $KOH_{(aq)}$. Stainless steel electrodes, emf = 5 V_{DC}, I = 0.5 A_{DC}, current density = 5 A/cm^2, room temperature. For color version of Fig. 8.47, see end sheets of book.

Composites

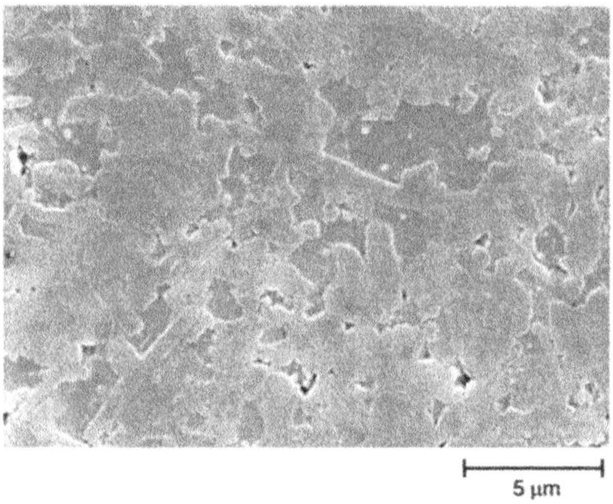

Fig. 8.49 Scanning electron microscope micrograph of SiC whisker-reinforced Al_2O_3 chemically etched 20 s in boiling phosphoric acid

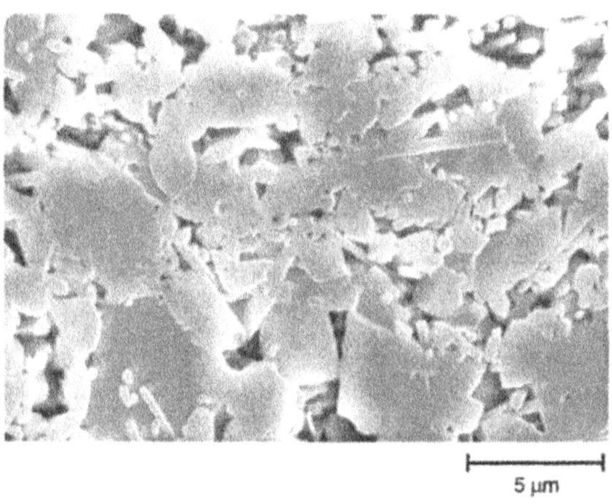

Fig. 8.50 Scanning electron microscope micrograph of SiC whisker-reinforced Al_2O_3 chemically etched 30 s in boiling phosphoric acid. Mean GS = 1.4 μm

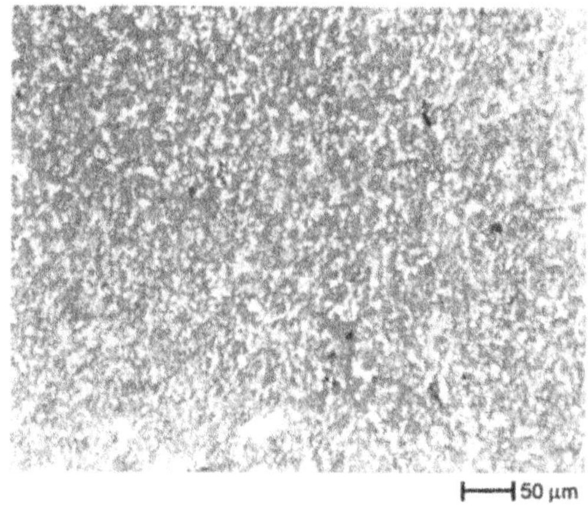

Fig. 8.51 Silicon carbide whisker-reinforced Al_2TiO_5 not etched. Viewed in differential interference contrast

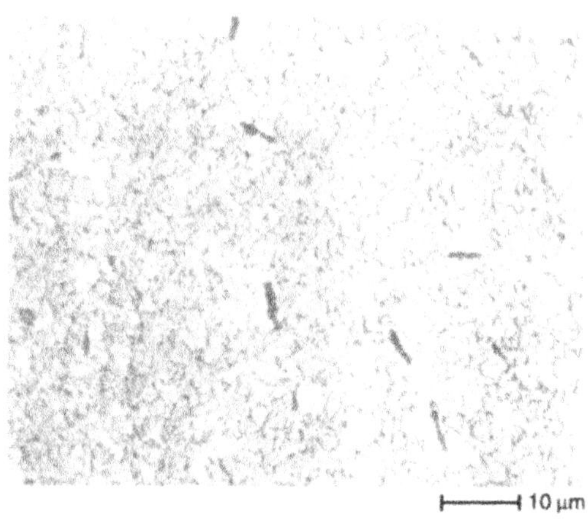

Fig. 8.52 Zirconia-toughened, hot pressed Al_2O_3 chemically etched 20 s in boiling phosphoric acid. Viewed in differential interference contrast

50 μm

Fig. 8.53 Mullite-alumina ($Al_6Si_2O_{13}$-Al_2O_3) refractory not etched

10 μm

Fig. 8.55 Silicon carbide (30 mass%) whisker-reinforced AlN. Not etched. Viewed in differential interference contrast

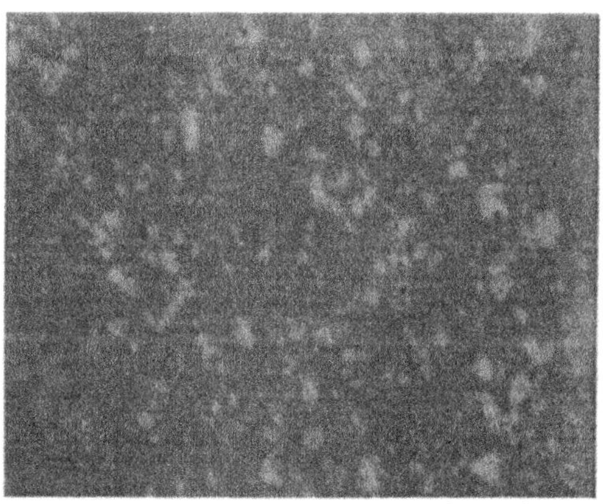

10 μm

Fig. 8.54 Silicon carbide whisker-reinforced Si_3N_4 not etched

10 μm

Fig. 8.56 Silicon carbide whisker-reinforced AlN. Not etched. Viewed in differential interference contrast

Metallized Ceramics

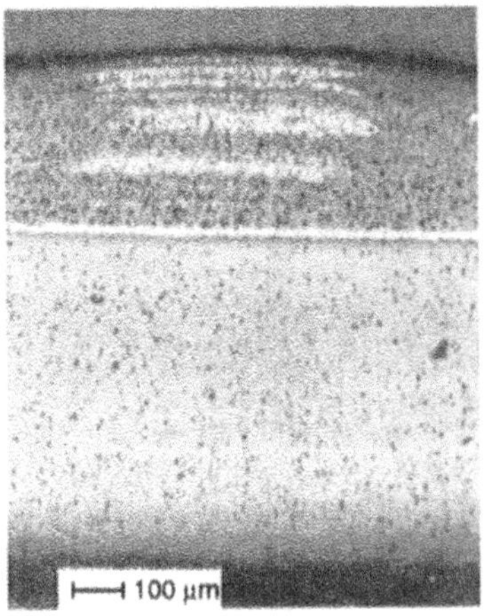

Fig. 8.57 Copper via stack (bright layers) in a dielectric ceramic on an Al_2O_3 multilayer interconnect board and not etched

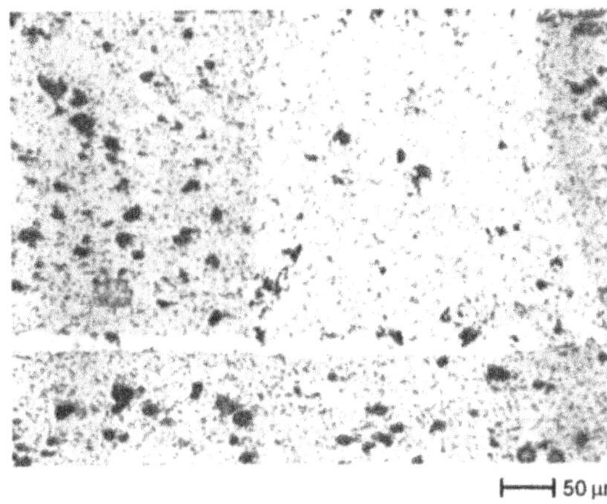

Fig. 8.59 Tungsten-metallized (bright phase) Al_2O_3 substrate. Not etched

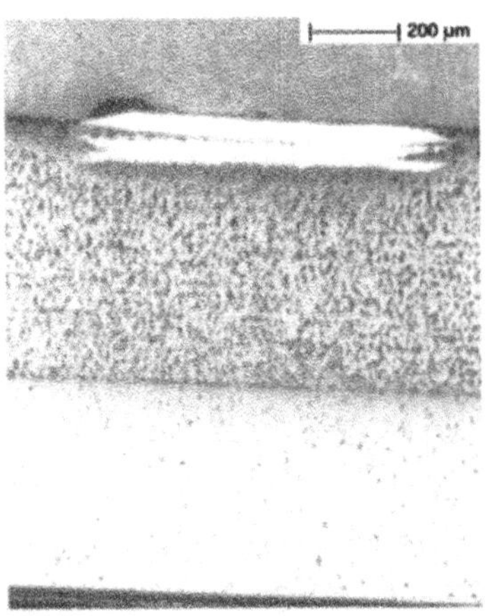

Fig. 8.58 Cross section of a multilayer interconnect board not etched. The bright phase on top is copper, the center porous phase is a dielectric ceramic, and the bottom phase is densified Al_2O_3.

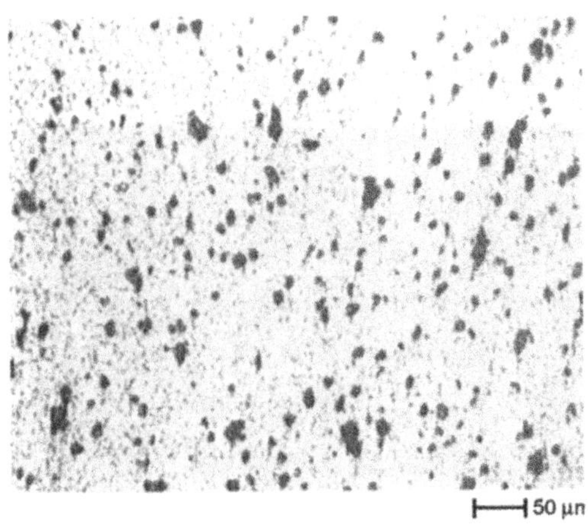

Fig. 8.60 Porous metallized Al_2O_3 substrate. Mean pore size = 7.3 μm, pore content = 7.5%, by image analysis

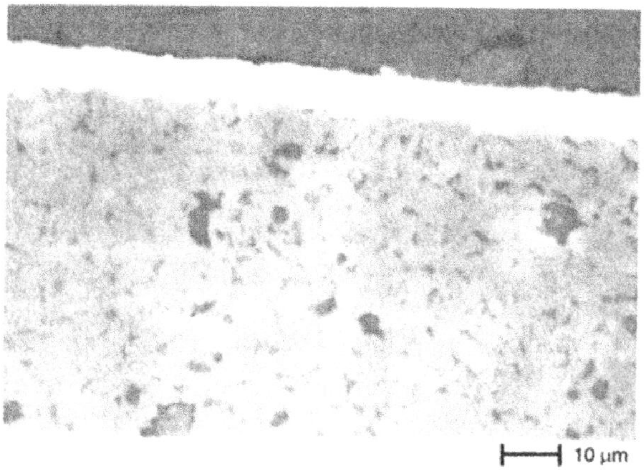

Fig. 8.61 Metallized (bright phase) Al_2O_3 substrate from Japan, not etched and mounted in castable acrylic (dark top layer)

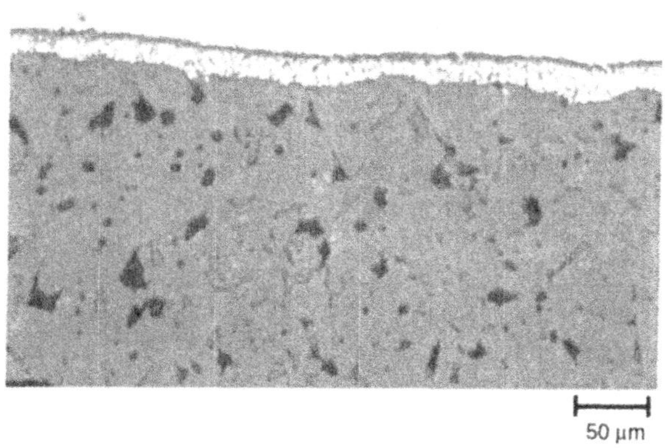

Fig. 8.62 Metallized (bright phase) 98% Al_2O_3 substrate, not etched and viewed through a green filter

Nitrides

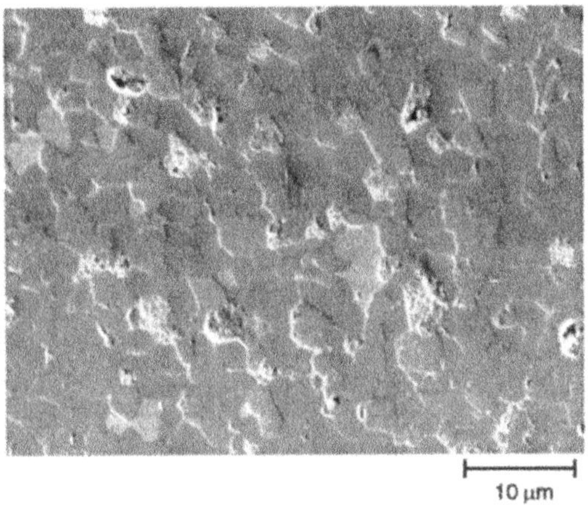

Fig. 8.63 Scanning electron microscope micrograph of AlN chemically etched 6 min in boiling nitric acid + glacial acetic acid

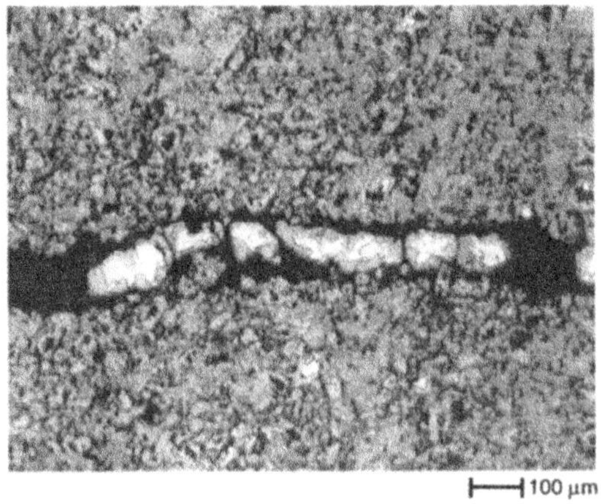

Fig. 8.65 Silicon nitride that fractured during sintering. Not etched

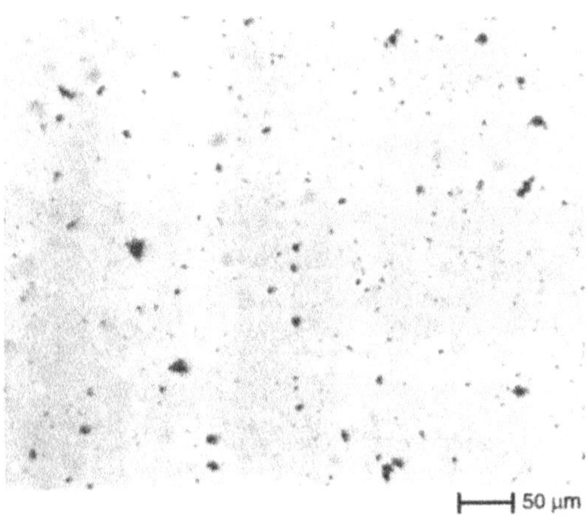

Fig. 8.64 Sialon, a solid solution of Si_3N_4 and Al_2O_3, not etched. Porosity (ASTM C 373, Archimedes' method) = 0.06%

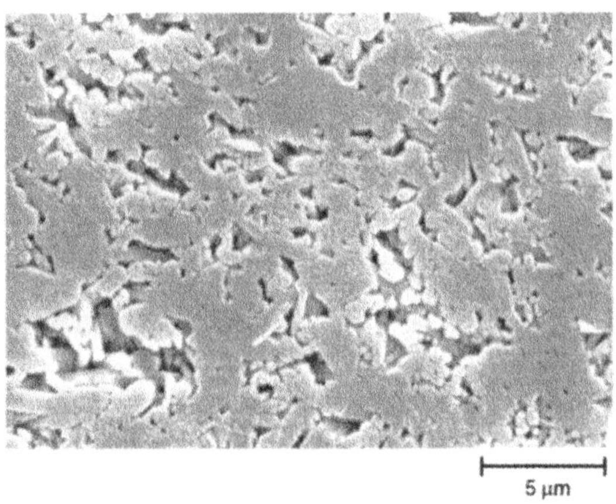

Fig. 8.66 Scanning electron microscope micrograph of Si_3N_4 chemically etched 15 min in boiling phosphoric acid. Mean GS = 0.57 μm

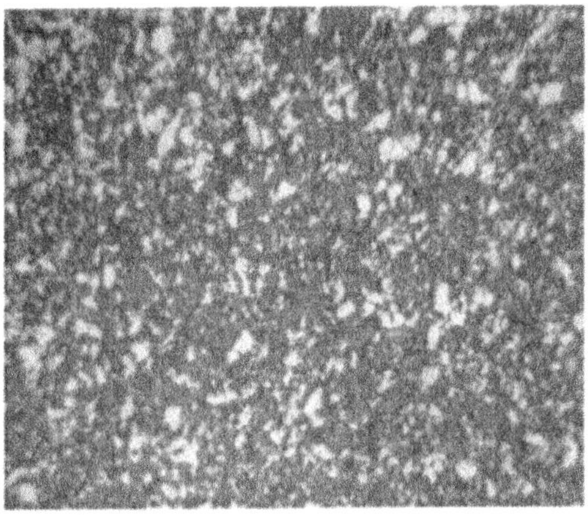

Fig. 8.67 Titanium nitride not etched

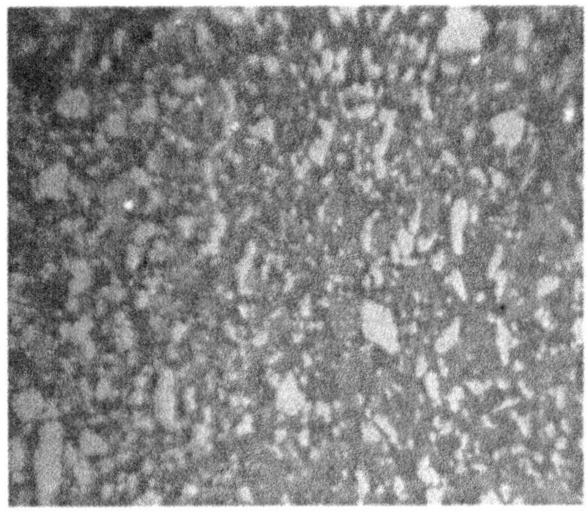

Fig. 8.68 Tialon, a solid solution of TiO_2 and AlN, not etched

Oxides

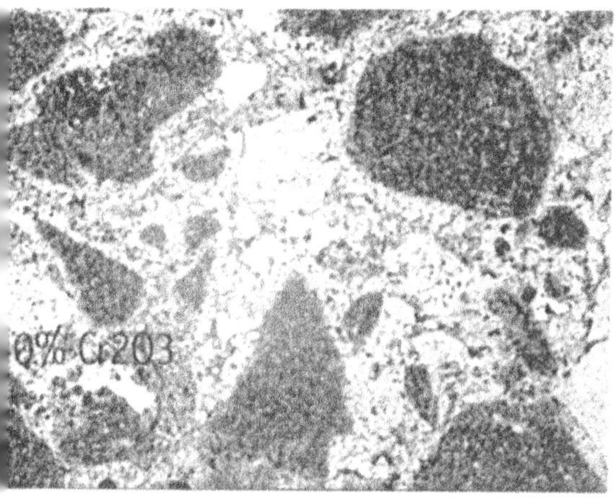

Fig. 8.69 40% Cr_2O_3 refractory brick not etched

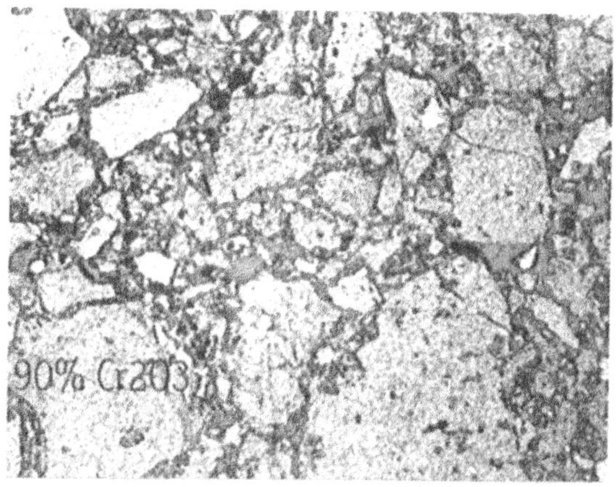

Fig. 8.71 90% Cr_2O_3 refractory brick not etched

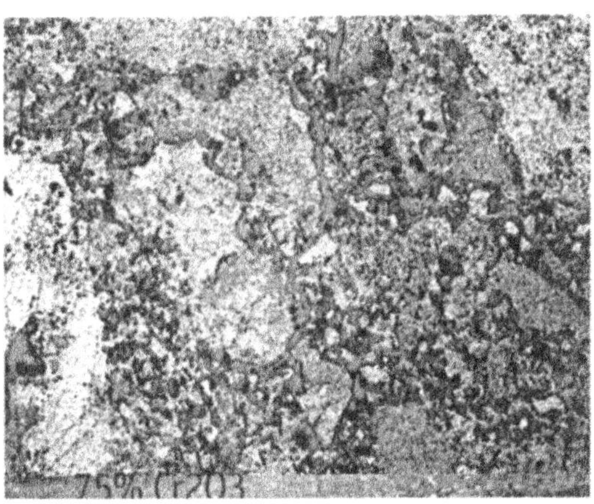

Fig. 8.70 75% Cr_2O_3 refractory brick not etched

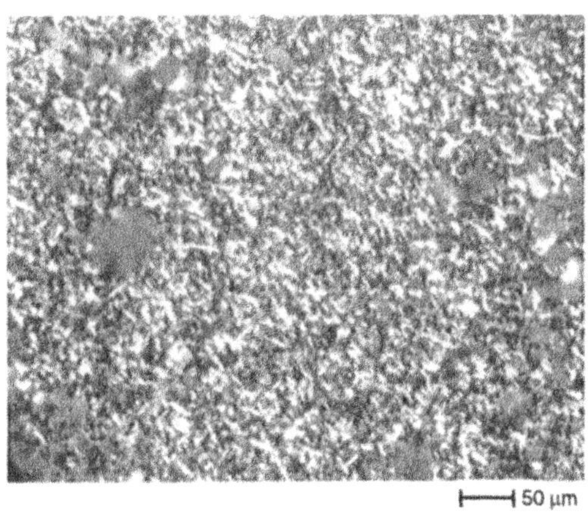

Fig. 8.72 Sillimanite ($Al_2O_3{\cdot}SiO_2$ or Al_2SiO_5) thin section

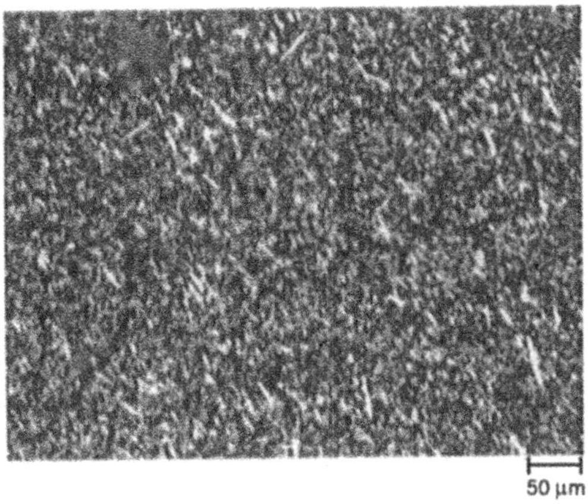

Fig. 8.73 Mullite ($3Al_2O_3{\cdot}2SiO_2$ or $Al_6Si_2O_{13}$) thin section

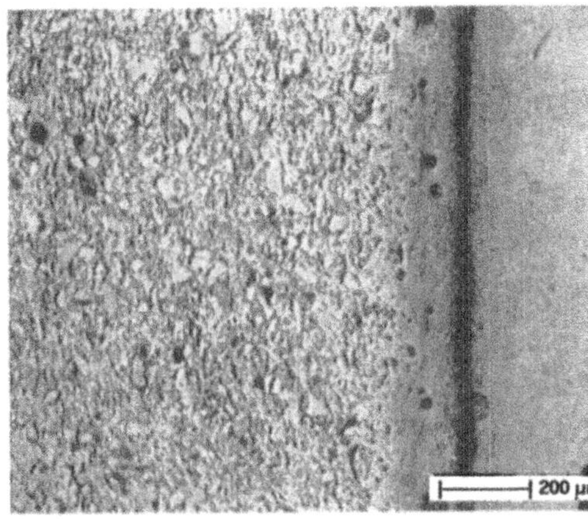

Fig. 8.75 Porcelain beverage mug cross section. The vitreous glaze is on the right side, above the scale bar.

Fig. 8.74 (a) Aluminum titanate ($Al_2O_3{\cdot}TiO_2$ or Al_2TiO_5) relief polished but not etched. Viewed in differential interference contrast. (b) Aluminum titanate ($Al_2O_3{\cdot}TiO_2$ or Al_2TiO_5) thin section

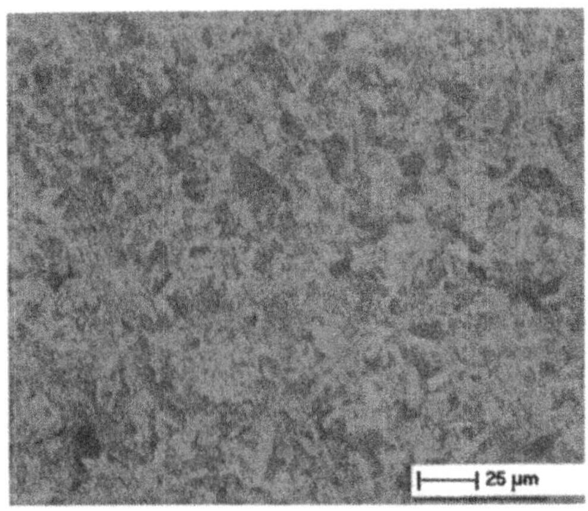

Fig. 8.76 $YBa_2Cu_3O_7$ superconductor etched 60 s in hydrofluoric acid at room temperature

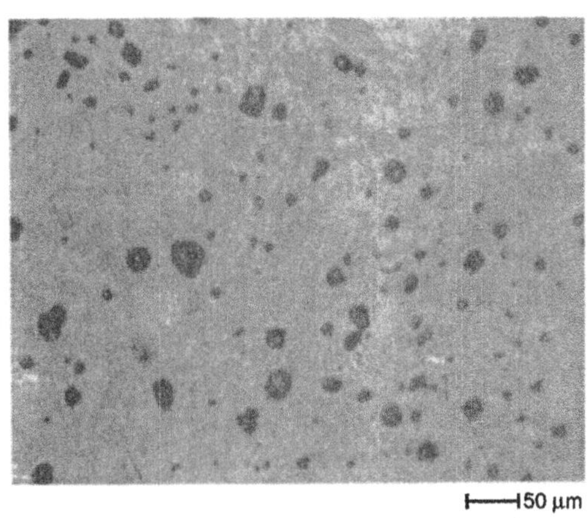

Fig. 8.77 Steatite ($MgSiO_3$) not etched

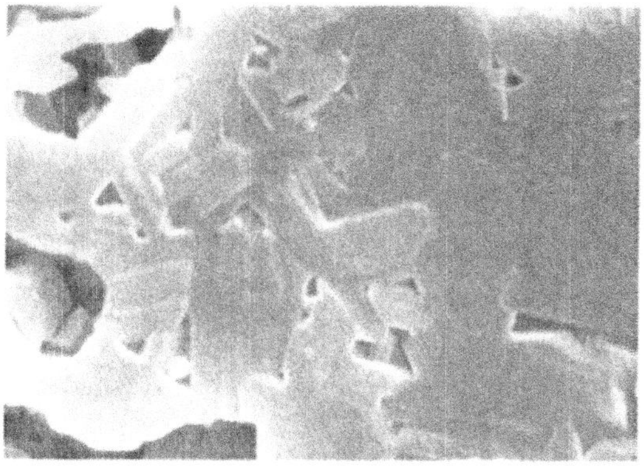

1 μm

Fig. 8.78 Scanning electron microscope photo of monoclinic $SrAl_2Si_2O_8$ (Ref 2) formed from Al_2O_3 dissolving into $SrO{\cdot}Al_2O_3{\cdot}SiO_2$ glass in a transient process (Ref 3). Chemically etched at room temperature for 20 s in dilute hydrofluoric + nitric acids

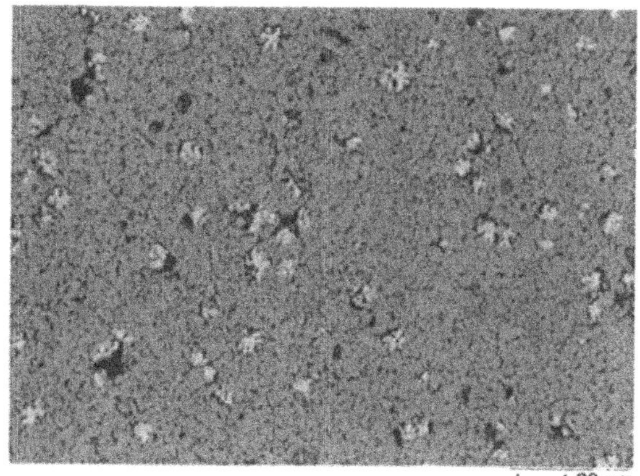

20 μm

Fig. 8.79 Scanning electron microscope backscattered electron photo of a zircon ($ZrSiO_4$) grinding pellet etched with dilute hydrofluoric acid and sputter coated with carbon. The white regions are ZrO_2.

Silicon Carbide

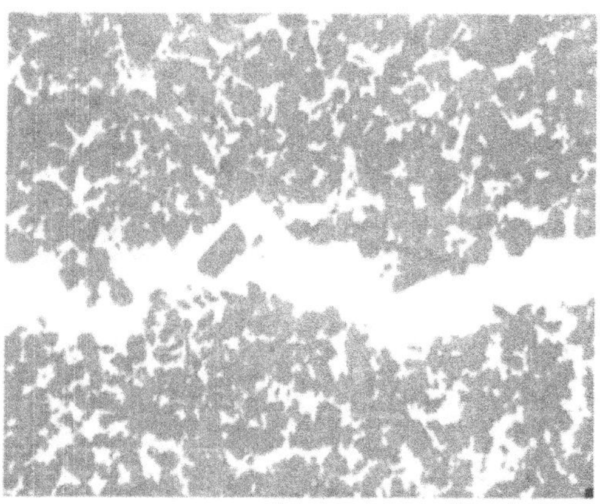

20 μm

Fig. 8.80 Reaction-bonded SiC that fractured during sintering. Molten free silicon filled the crack. Not etched

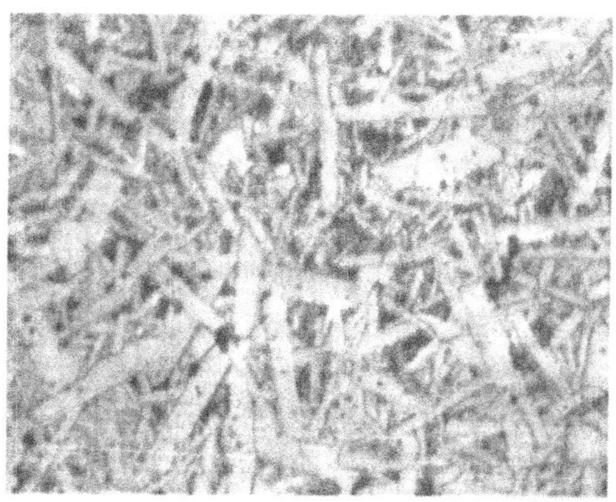

10 μm

Fig. 8.81 Direct-sintered SiC chemically etched 6 min in boiling Murakami's etchant. Optical filters and polarization were not used in the photograph. Mean GS = 2.3 μm

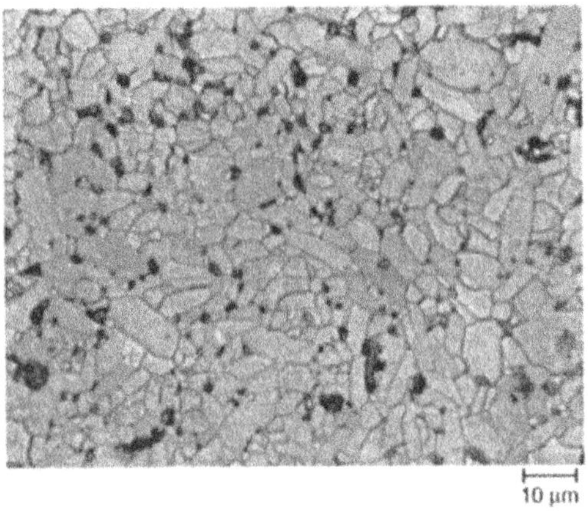

Fig. 8.82 Direct-sintered SiC chemically etched 8 min in boiling Murakami's etchant. Mean GS = 3.8 µm

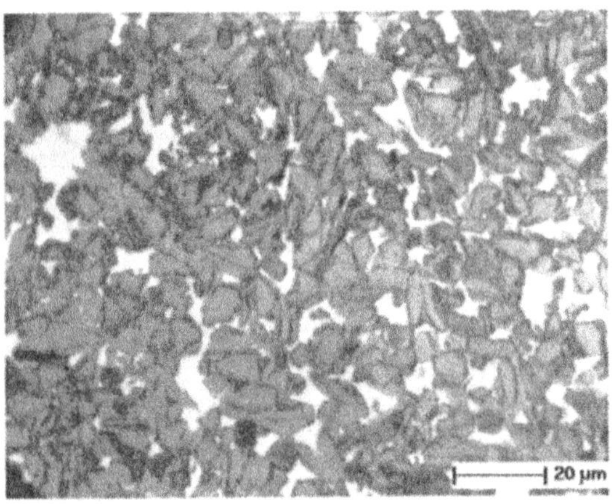

Fig. 8.83 Reaction-bonded SiC electrolytically etched in aqueous 0.36 *M* KOH for 3 min at 6 V_{DC} and 2 A on a 4 × 9 mm polished section. Mean GS = 4.0 µm

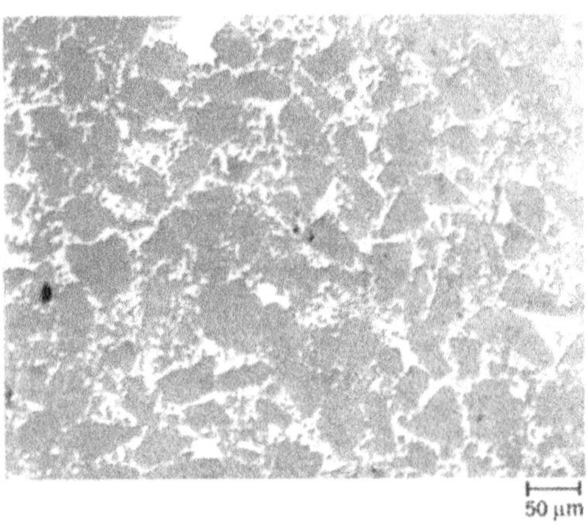

Fig. 8.84 Reaction-bonded SiC not etched. The light phase is free silicon.

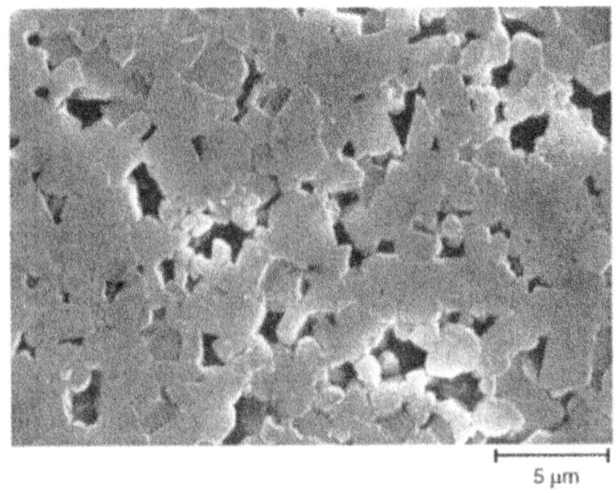

Fig. 8.85 Direct-sintered SiC, toughened with glass, from Japan. Chemically etched in boiling Murakami's etchant. Mean GS = 1.2 µm

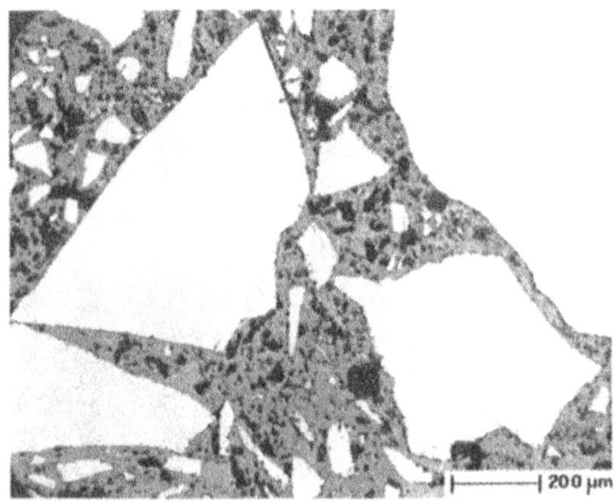

Fig. 8.86 Silicon-nitride-bonded SiC refractory. Not etched. The large, light grains are SiC.

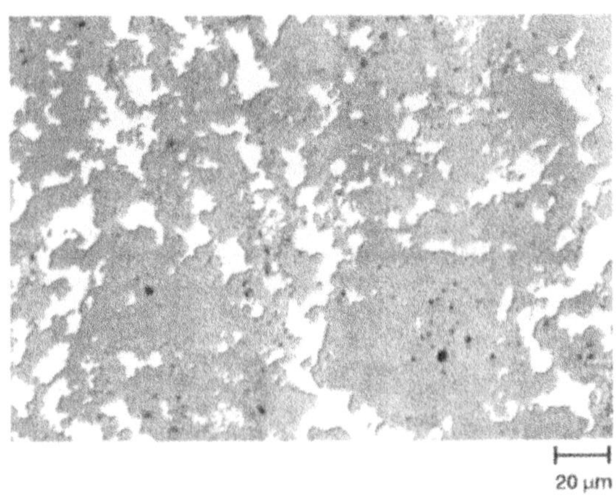

Fig. 8.87 Reaction-bonded SiC not etched. Relief-polished and viewed in differential interference contrast. The light phase is free silicon.

Spinel

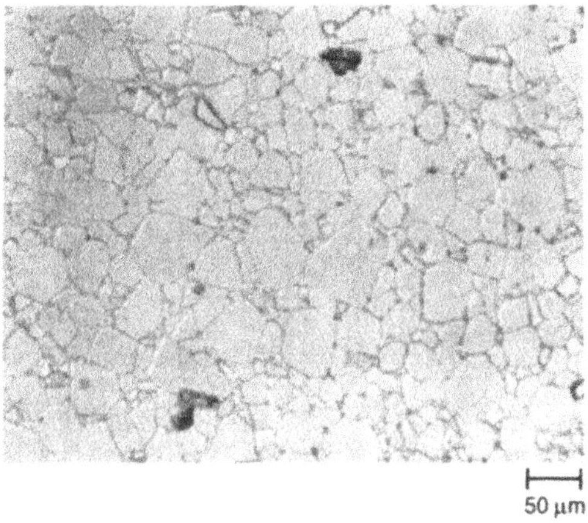

Fig. 8.88 $MgAl_2O_4$ thermally etched and gold sputter coated. The mean grain size is 19 μm by the circle-intercept method, 20 μm by the lineal intersection method, and 19 μm by image analysis (Ref 1).

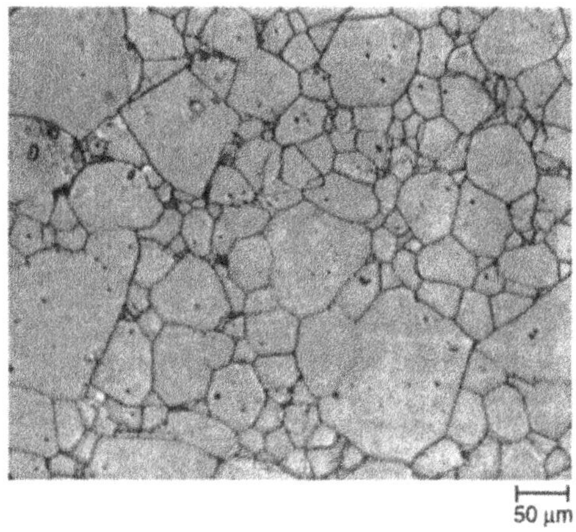

Fig. 8.90 Hot isostatically pressed $MgAl_2O_4$ thermally etched 45 min at 1500 °C. Mean GS = 32 μm

Fig. 8.89 $MgAl_2O_4$ thermally etched 60 min at 1500 °C and gold sputter coated

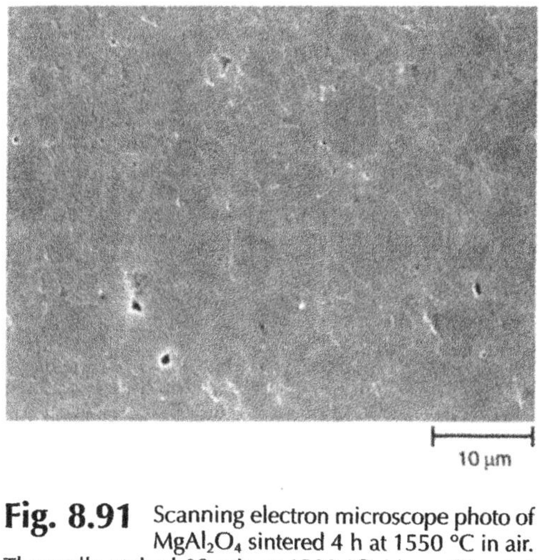

Fig. 8.91 Scanning electron microscope photo of $MgAl_2O_4$ sintered 4 h at 1550 °C in air. Thermally etched 60 min at 1500 °C. Mean GS = 1.4 μm

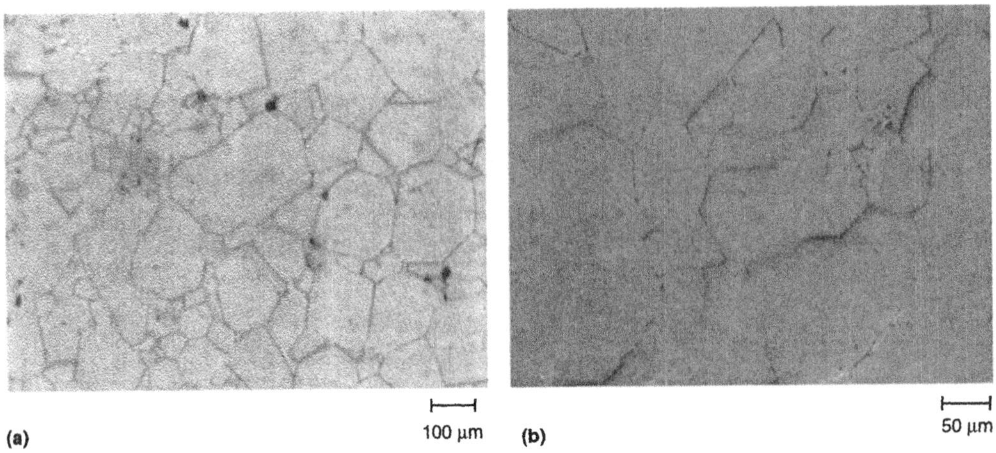

Fig. 8.92 (a) Hot isostatically pressed $MgAl_2O_4$ relief polished and thermally etched 30 min at 1500 °C. The original Polaroid photo was exposed for 10 s. Mean GS = 72 μm. (b) Same hot isostatically pressed $MgAl_2O_4$ as in (a) but at twice the magnification and viewed in differential interference contrast. The original Polaroid photo was exposed for 75 s. Mean GS = 72 μm

Zirconia

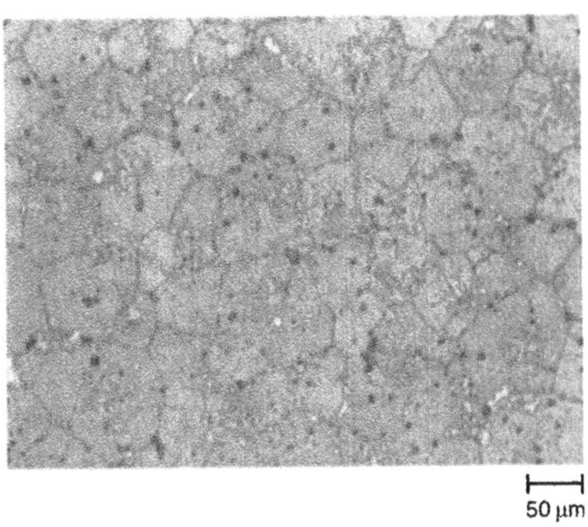

Fig. 8.93 Transformation-toughened ZrO_2 partially stabilized with MgO. Chemically etched 60 s in a boiling aqueous solution of NH_4HF_2 + HF + H_2O_2

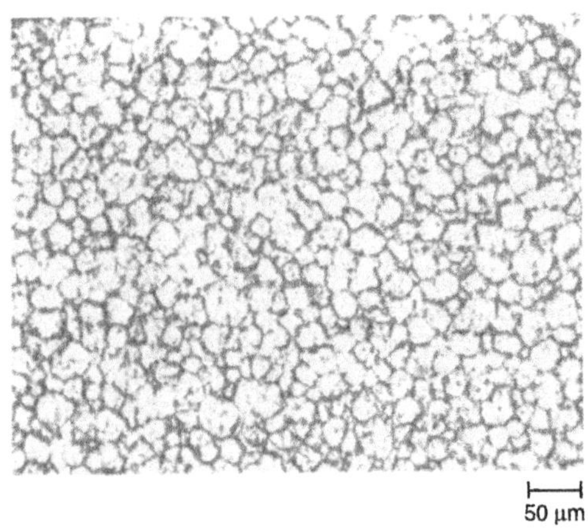

Fig. 8.94 Zirconia thin section partially stabilized with MgO. The primary (white) phase is cubic. The grain boundaries are marked by tetragonal and monoclinic phases.

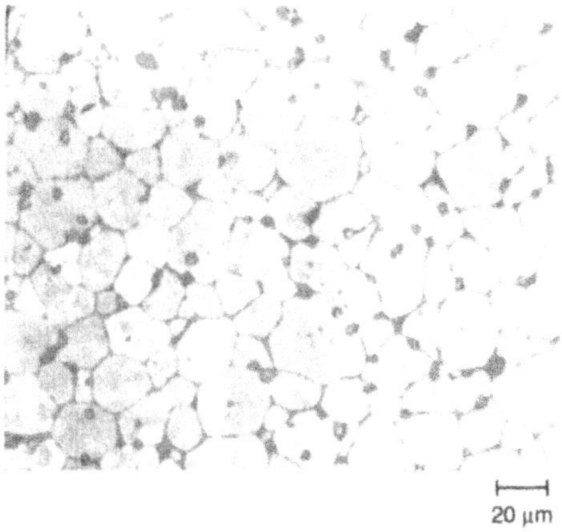

Fig. 8.95 Tetragonal ZrO_2 polycrystals. Chemically etched 50 s in a boiling aqueous solution of NH_4HF_2 + HF + H_2O_2. Mean GS = 13 μm

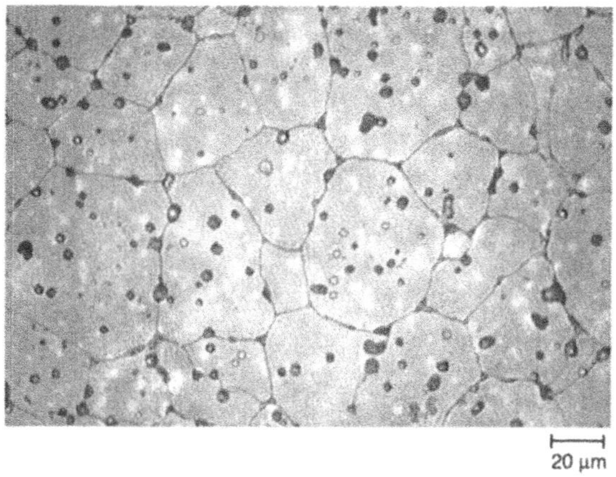

Fig. 8.97 Cubic ZrO_2 fully stabilized with Y_2O_3. Thermally etched 30 min at 1475 °C in air

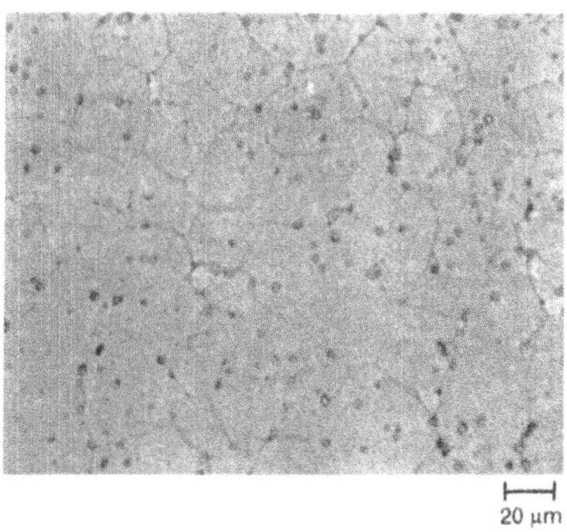

Fig. 8.96 Cubic ZrO_2 fully stabilized with Y_2O_3. Chemically etched 60 s in a boiling aqueous solution of NH_4HF_2 + HF + H_2O_2

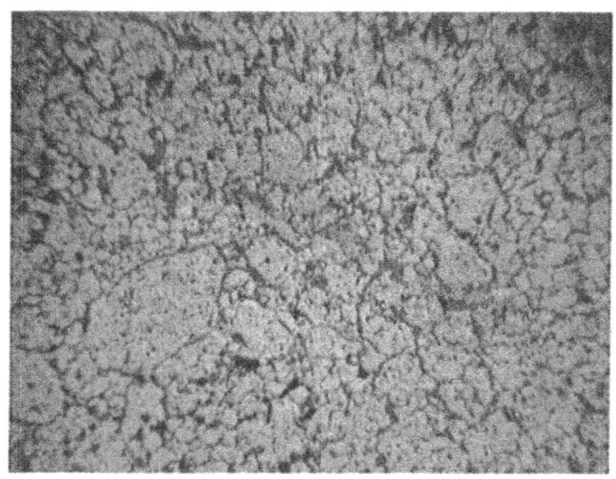

Fig. 8.98 Zirconia refractory not etched

REFERENCES

1. R.E. Chinn, Grain Sizes of Ceramics by Automatic Image Analysis, *J. Am. Ceram. Soc.*, Vol 77 (No. 2), 1994, p 589–592
2. R.E. Chinn, "Transient Glass-Phase Processing of $SrAl_2Si_2O_8$-Alumina Ceramics," Thesis T-4685, Colorado School of Mines, 1995, p 90, 108
3. R.E. Chinn, M.J. Haun, C.Y. Kim, and D.B. Price, Low-Temperature Transient Glass-Phase Processing of Monoclinic $SrAl_2Si_2O_8$, *J. Am. Ceram. Soc.*, Vol 81 (No. 9), 1998, p 2285–2293

CHAPTER 9

Quantitative Ceramography

SEVERAL QUANTITATIVE measurements of microstructures are demonstrated in this chapter. Ceramography is used to quantify the grain size, grain shape, porosity, second-phase content, microindentation hardness, and toughness of ceramics. Many bulk mechanical, thermal, electrical, magnetic, and optical properties of ceramics are related to microstructural quantities.

Stereology

Stereology is the science of quantification or characterization of three-dimensional objects by measurement of two-dimensional cross sections or projections. Quantitative measurements of microstructures, such as grain size and pore size, depend on stereographic principles. One of the cornerstones of stereology is that the volume fraction (V_V) of a phase in a ceramographic cross section is equal to its area fraction (A_A), lineal fraction (L_L), and point fraction (P_P), as expressed in Eq 1 (Ref 1):

$$V_V = A_A = L_L = P_P \qquad \text{(Eq 1)}$$

Several techniques discussed in this chapter use the area fraction, lineal fraction, and point fraction to measure grain size, second-phase content, and porosity in ceramics. Russ and Dehoff's book explains stereology in great detail, along with some caveats regarding grain size, grain shape, and second-phase content measurements.

Grain Size

The mean *diameter* of the grains in a microstructure can be defined in several ways. For example, the grain size could be taken as the diameter of a circle with the same area as the mean grain area in a ceramographic cross section. Ideally, this would be equivalent to the diameter of a sphere with the same volume as the mean grain volume. The cross section cuts through grains randomly, such that what appears to be a very small grain in the cross section could actually be the corner or tip of a much larger grain. Only a few of the hundreds of grains in a typical micrograph are likely to be bisected by the plane of the photo through their largest cross section, such that the "true" grain size is somewhat larger than what is measured in the cross section. Even if all the grains were spheres, the mean intercept length would only be two-thirds of the diameter (Ref 1). Some stereological characterizations of particle diameter are shown in Fig. 9.1.

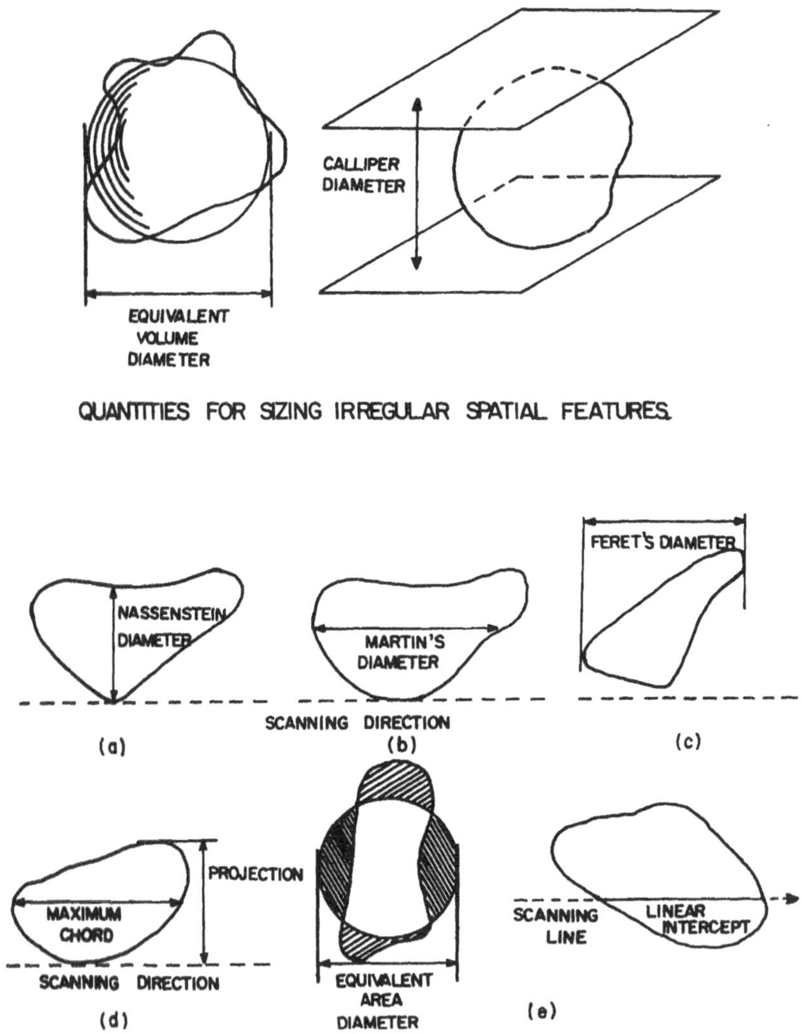

Fig. 9.1 Some stereological definitions of particle diameter (Ref 2)

The grain size and other microstructural characteristics are all dependent on the processing conditions and methods used to fabricate the ceramic. Inversely, these microstructural measurements are indicators of the how the ceramic was processed. For example, a grain size that exceeds specifications could indicate that the sintering temperature was too high, the sintering time too lengthy, or the initial particle size too small.

The grain size affects a wide variety of mechanical, thermal, optical, and electrical *bulk* properties of ceramics and also indicates the extent of sintering. The bulk properties are the characteristics of an entire piece, which are affected by grains, pores, and flaws, as opposed to the properties of a single crystal that are dependent mostly on interatomic bonding. A general and somewhat oversimplified but very common correlation between the grain size and a bulk mechanical property was first proposed by Hall in 1951 and modified by Petch in 1953 as Eq 2:

$$\sigma_f = \sigma_0 + \frac{k_1}{\sqrt{d}} \qquad \text{(Eq 2)}$$

where σ_f is fracture stress; σ_0 is friction stress, a measure of lattice resistance to deformation not necessarily equal to the strength of a single crystal; k_1 is a constant with respect to grain size but a function of flaw surface energy, elastic modulus, and Poisson's ratio, with the same dimensions as toughness; and d is the equivalent grain diameter.

The Hall-Petch equation is discussed in greater detail in Ref 3, 4, and especially 5. The Hall-Petch parameters are constant only within certain grain size ranges—fine-grain ceramics tend to have a low k_1 (the slope on the plot in Fig. 9.2) and a high σ_0 (intercept), while coarse-grain ceramics tend to have a high k_1 and a low σ_0. The slope k_1 in Eq 2 has the same dimensions as toughness (see the section "Toughness" in this chapter) and may be related. The transition region between the two grain size ranges, where the failure mode changes from microplasticity to the Griffith flaw mechanism, is usually where half the grain size is approximately equivalent to the critical flaw size. The parameters of Eq 2 depend on many complex characteristics of the ceramic and its microstructure, such as anisotropy, surface finish, porosity and pore size, grain shape, test specimen size, and multiple phases. If the failure originates at an oversize grain, the maximum grain size may be more useful than the mean grain size, as d in Eq 2 (Ref 5).

The gist of the Hall-Petch equation is that the strength of a ceramic decreases nonlinearly as its grain size increases. Thus, an overfired ceramic is likely to fail at an unexpectedly low stress, because of its oversize grains. Ceramic manufacturers frequently use grain size as a quality-control measure for this reason. Rice (Ref 5) and many others since the late 1940s have developed mathematical models correlating strength, hardness, wear, toughness, thermal shock, dielectric constant, and other properties to grain size and porosity for various ceramic compositions.

The mean grain diameter is measured by one of five methods: planimetric, line-intercept, circle-intercept, comparison, or image analysis. A specimen measured by all five methods is likely to have five different grain sizes, because the methods do not measure the exact same thing.

Planimetric Method. The planimetric method, devised by Jeffries et al. in 1917, is described in detail in section 8 of ASTM E 112. This method determines mean grain size from the number of grains in a field of view or micrograph that lie, totally and partially, within a circle or rectangle of known area. From a table in ASTM E 112, the number of grains per unit area is converted to a nominal diameter. The accuracy and applicability of this method are affected by secondary phases, such as glass and porosity.

Lineal-Intercept Method. The lineal-intercept method devised by Heyn in 1903 is described in section 10 of ASTM E 112. This method determines mean grain size from the number of grains (intercepts) or grain

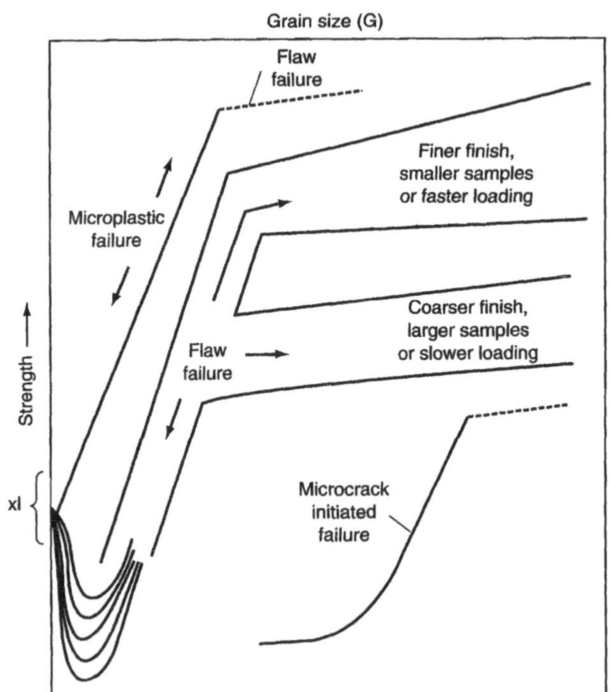

Fig. 9.2 Schematic representation of the strength of a ceramic as a function of the inverse square root of grain size in the preexisting-flaw model. The slope changes abruptly where the critical flaw size is approximately half the grain size. The "xl" intercept on the ordinate indicates the relative strength of a single crystal (Ref 5).

boundaries (intersections) that intersect one or more lines of known length superimposed on a field of view or micrograph. Nonequiaxed and oriented microstructures require the line to be superimposed in several random directions to compensate for the nonuniformity of individual grains. Equation 3 correlates the equivalent grain diameter to the intercept count:

$$d = \frac{L \times A_1}{M \times N} \qquad \text{(Eq 3)}$$

where d is the equivalent grain diameter; L is the length of superimposed line(s); A_1 is the shape correction factor, nominally 1.00; M is magnification (dimensionless); and N is the quantity of intercepts or intersections.

Mendelson (Ref 6) used 1.56 as the correction factor for ceramics. Wurst and Nelson (Ref 7) modified Eq 3 to take into account secondary phases. The median of the observed diameters is 0.87 of the true diameter of spherical grains, and the mean diameter is 0.82 of the true diameter, by statistical inference (Ref 8). Therefore, 1.22 (the reciprocal of 0.82) could be used as the correction factor for equiaxed grains in Eq 3.

Circular-Intercept Method. The circular-intercept procedures devised by Hilliard in 1964, Underwood and Coons in 1965, and Abrams in 1971 are similar to Heyn's method except that circles of known circumference are used instead of lines. Circles compensate for nonequiaxed and oriented microstructures and intercepts at the ends of test lines. These methods, described in ASTM E 112, are affected strongly by secondary phases and slightly by the curvature of the circle as it crosses a grain. Equation 4 correlates the equivalent grain diameter to the intercept count on a circle:

$$d = \frac{C \times A_1}{M \times N} \qquad \text{(Eq 4)}$$

where C is the circumference of the superimposed circle, and the other parameters are the same as in Eq 3.

Example 9.1: Estimate the mean grain size of the spinel in Fig. 8.90 by the lineal intersection method.

Solution. From Eq 3, assume by default $A_1 = 1.0$. The sides of the square in Fig. 9.3(a) are 294 μm long (*L/M*). The quantity of intersections along all four sides is approximately 44. The triple points (where three grains intersect) are counted as 1.5, and the line ends are 1/2 where they do not intersect a grain boundary. The left end of the bottom line is right on an intersection, but the right end is inside a grain:

$$d = \bar{l} = \frac{4(294\ \mu\text{m})(1.0)}{44} \approx 27\ \mu\text{m}$$

The small quantity of intersections, combined with the wide range of grain sizes, results in a large standard deviation when several fields are counted. Notice that the large grain that intersects the lower right corner of the square has approximately ten times as much area as most of its contiguous neighbors and upward of 100 times the area of the smallest grains in the image.

Example 9.2: Estimate the mean grain size of the spinel in Fig. 8.90 by the planimetric method.

Solution. In Fig. 9.3(a), the quantity of grains inside the square is approximately 101.5. The grains entirely within the square count as one, and the grains that intersect the square count as 1/2. Because the spinel is transparent similar to window glass, some subsurface grain boundaries are faintly visible and can be misleading. The left side of the triangular grain near the center of the upper half of the square is an example of this. The very smallest grains are also difficult to count, and so, any two microscopists are unlikely to count exactly the same number of grains. These "errors" would be compensated for by evaluating several different photos of the same specimen and calculating the standard deviation, coefficient of variation, and confidence limit. The mean area is equal to the area of the square divided by the quantity of grains:

$$\bar{a} = \frac{(294\ \mu\text{m})^2}{101.5} = 852\ \mu\text{m}^2 = 0.852 \times 10^{-3}\ \mu\text{m}^2$$

The intercept size is assumed to be equivalent to the diameter of a circle of the same mean area. The spinel is more or less equiaxed, so the assumption is valid in this case. The assumption would be less valid for non-

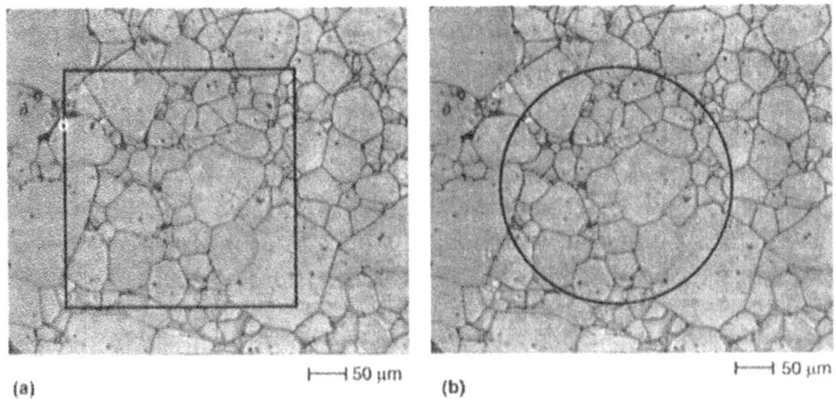

Fig. 9.3 Circular-intercept method for determining grain size. (a) Superimposed lines on Fig. 8.90 for counting grain-boundary intersections. (b) Superimposed circle on Fig. 8.90 for counting grain intercepts

equiaxed grains, such as most of the 96% alumina sections shown in the section "Aluminas" in Chapter 8:

$$\bar{l} = d_{eqv} = \sqrt{\frac{4\bar{a}}{\pi}} = \sqrt{\frac{4(852\ \mu m^2)}{\pi}} = 33\ \mu m$$

From Table 2 in ASTM E 112 or Eq 5(b), the ASTM micrograin size number G is 7.0 to 7.5.

Example 9.3: Estimate the mean grain size of the spinel in Fig. 8.90 by the circular-intercept method.

Solution. From Eq 4, assume by default $A_1 = 1.0$. The diameter of the circle in Fig. 9.3(b) is 286 μm, so its resolved circumference C/M is 900 μm. Going clockwise from the tick mark at the 12:00 position, there are approximately 27 intercepts. The triple points (where three grains intersect) are counted as two intercepts in the Hilliard method, such as those near the 7:00 and 11:00 positions. The surface is not entirely opaque, and some subsurface grain boundaries are faintly visible, such as those between 5:00 and 6:00. The subsurface boundaries are not counted:

$$d = \bar{l} = \frac{(900\ \mu m)(1.0)}{27} \approx 33\ \mu m$$

Normally, a 250 mm circumference circle photocopied on an acetate overlay is superimposed on a micrograph of standard magnification (200, 500, 1000×, etc.), which allows use of the nomographs in ASTM E 112. For reporting purposes, three or more representative micrographs of the specimen should be counted and averaged.

Image Analysis. The image analysis method, described in greater detail in Chapter 10, uses a video camera and computer to digitize the image of a microstructure. In the digitization step, the computer reads the image as an array of *pixels* (picture elements) that are characterized with *x*-, *y*-, and sometimes *z*-coordinates and a gray level. Prior to measurement, the computer binarizes the digital image into white (gray level = 255) objects—the primary grains—and black (gray level = 0) background. The background consists of grain boundaries, pores, secondary phases, and artifacts. The computer measures the binarized grains individually, thus distinguished from the remainder of the microstructure, and statistically compiles the measurements.

In this way, the image analyzer can measure many more parameters than just grain size, as discussed in subsequent sections. The advantages of image analysis compared to the other methods include its:

- Ability to measure all the grains in a field of view as opposed to just the grains that lie along a line or circle

- Adaptability to automation for quick, reproducible, unbiased results
- Ability to measure grains separate from other microstructural features

The disadvantages of image analysis include:

- Some microstructures are not easily binarized, especially ceramics, because the gray levels or red, green, blue values of various features tend to overlap.
- Some grain boundaries, invariably, are discontinuous or obscured.
- Image analyzers are expensive and initially require the skills of an experienced programmer.
- Errors caused by such factors as bad calibration may be difficult to detect.

Semiautomatic image analyzers that require the operator to trace grains onto an overlay with an electronic stylus or mouse also have the disadvantage of human bias and, perhaps, sloppy artwork.

Two ASTM procedures that specifically apply to image analysis, E 1245 and E 1382, are listed in Appendix A.

Comparison Method. The comparison method of grain size evaluation is simply the comparison of a micrograph of the microstructure in question to a micrograph at the same magnification of a very similar material of known grain size. The standard micrograph is measured by one of the other methods. Because this method is not a true measurement, it precludes quantified accuracy and statistical characterizations, such as standard deviation. The advantage of this method is that it is fast and easy and well suited to situations where a "ball-park figure" is acceptable.

One application for the comparison method is in thin-section evaluation. Thin sections are not easily quantified by the other methods because many grains overlap, the smaller grains are obscured, and the "plane" of focus is distorted in the *z*-dimension. If an alumina section is polished, etched, and quantified by one of the other methods, it can be made into a thin section and used as a standard for other components of similar composition.

It is important that a 96% alumina section, for example, is compared only to another 96% alumina and not to an 85, 94, or 99.5% alumina. Furthermore, if the 96% alumina was uniaxially pressed, it is best compared to a standard that was fabricated by the same method rather than, say, an isostatically pressed 96% alumina. The minor components and the fabrication method as well as the initial particle size and the sintering conditions are likely to have an effect on the final grain size and shape.

ASTM Grain Size Number. The ASTM grain size number G is a logarithmic representation of the grain size based on the quantity of grains within a unit area. The grain size number has an inverse relationship to the mean diameter or intercept length; that is, the smallest grains have the largest grain size number. Micrograin size No. 1 has 1.0 grain/in.^{2}, or 15.5 grains per 10,000 mm^2, at 100× magnification. Seldom used in the ceramics industry, the ASTM micrograin size number is determined from the tables in ASTM E 112 or Eq 5(a) and (b):

$$G = -6.65\log\left(\frac{L}{N}\right) - 3.3 \qquad \text{(Eq 5a)}$$

$$G = -3.32\log\left(\frac{N}{A}\right) - 2.95 \qquad \text{(Eq 5b)}$$

In Eq 5(a) and (b), N is the quantity of intersections or intercepts, L is the actual length of the test line or circumference of the test circle in millimeters, and A is the actual area of the test rectangle or circle in square millimeters. The ASTM grain size number is usually expressed to the nearest half-integer.

Grain Size Distribution. Grain sizes in a ceramographic section are distributed around at least one mean value, often very widely, for at least three reasons:

- The "diameter" varies substantially within each nonequiaxed grain, where the maximum diameter might be an order of magnitude greater than the minimum.
- The plane of polish intersects grains at random, sometimes through their largest dimension and sometimes across a corner or tip.
- The actual size varies among neighboring grains in a ceramic, especially in high-purity compositions such as 99.8% alumina, which often lack sintering aids that control grain growth.

Standard deviation is a measure of the dispersion of grain sizes around the mean and generally incorporates all of the previously mentioned factors. The third factor, true dispersion, is best characterized by grain size measurements in three mutually perpendicular cross sections.

Some ceramics have duplex or bimodal microstructures where few, if any, grains are the size of the mean diameter. Figures 8.31 and 8.35 show duplex 99.5% alumina microstructures where discrete larger grains are surrounded ("necklaced") by a matrix of smaller grains. In this case, the mean diameter is misleading, because the microstructure really has two means approximately one order of magnitude apart. The microstructure contains many more small grains than large ones, resulting in an exponential distribution of grain sizes. ASTM E 1181, listed in Appendix A, provides methods for characterizing duplex microstructures.

Example 9.4: A cold isostatically pressed 99.5% alumina microstructure was photographed in ten regions at 500× magnification. A 250 mm circumference (79.6 mm diameter) circle was superimposed on each photo, and the intercepts were counted. The counts were N_i = 59, 62, 54, 58, 55, 61, 74, 68, 66, and 46. How would you characterize the grain size distribution of this specimen?

Solution: First, calculate the grain size in each photo from Eq 4, with the circumference multiplied by 1000 to convert millimeters to microns:

$$d_i = \frac{(250\text{ mm})(1000\frac{\mu\text{m}}{\text{mm}})}{500N_i} = 8.5, 8.1, 9.3, 8.6, 9.1, 8.2, 6.8, 7.4, 7.6, \text{ and } 10.9\ \mu\text{m for the ten photos}$$

The mean grain size, $\bar{d}$, the arithmetic mean of the previously mentioned ten calculations, is 8.4 μm.

Second, calculate the standard deviation, s, of the grain size, a statistical description of the dispersion of data about the mean. The standard deviation is best calculated with a spreadsheet or a scientific calculator with built-in statistical functions. The standard deviation equation is:

$$s = \sqrt{\frac{\sum_{i=1}^{N}(d_i - \bar{d})^2}{N-1}} = 1.1\ \mu\text{m}$$

The N term in the standard deviation equation is the number of measurements or samples in the set, ten.

Third, calculate the coefficient of variation, CV. The coefficient of variation is the ratio of the standard deviation to the mean, usually expressed as a percentage:

$$CV = \frac{s}{\bar{d}} = \frac{(1.1)(100\%)}{8.4} = 13\%$$

The intercept counts, N_i, also could have been used to calculate the standard deviation and coefficient of variation. From Fig. 9 in ASTM E 112, the relative 95% confidence limit for 10 tests and $CV = 0.13$ is 0.09, or 9%. The 95% confidence interval for the grain size is:

$$CI_{0.95} = 8.4\ \mu\text{m} \pm (0.09)(8.4\ \mu\text{m}) = 8.4\ \mu\text{m} \pm 0.76\ \mu\text{m} = \{7.7\ \mu\text{m}, 9.2\ \mu\text{m}\}$$

One can be confident that 95% of the grains are at least 7.7 μm in diameter and no more than 9.2 μm. The ASTM grain size number, seldom used in the ceramics industry, is determined from Fig. 6 in ASTM E 112. The arithmetic mean of the ten counts is approximately 60. If it is assumed that twice as many grains would have been counted on a 500 mm test pattern, then the mean count would be 120. The ASTM micrograin size number corresponding to the intersection of $N = 120$ and magnification = 500× is approximately 10.5. Alternatively, from Eq 5(a):

$$\bar{N} = \frac{\Sigma N}{10} = 60.3$$

$$L = \left(\frac{250\text{ mm}}{500}\right) = 0.5\text{ mm}$$

$$G = -\,6.65\log\left(\frac{0.5}{60.3}\right) - 3.3 = 10.5$$

Grain Shape

The shapes of grains can be statistically quantified by dimensionless ratios called *shape factors, form factors,* or *shape indexes.* The shape factor ratios typically include axial lengths, areas, perimeters, and moments of grain shapes. Properties that are sensitive to grain shape, such as the toughness of ceramics that fracture intergranularly and the effectiveness of abrasive particles, are sometimes correlated to shape factors. Grain shape may have some effect on strength and fracture mode (Ref 5).

Shape factors should be used with caution, because two microstructures that look quite different could have similar shape factors. Furthermore, two-dimensional shape factors do not necessarily correspond well to actual grain shapes in three dimensions.

Aspect ratio is the ratio of the largest diameter of a grain to the smallest, as in Eq 6(a). For a very equiaxed grain, the aspect ratio would be close to one. For a very oblong grain such as a whisker, the aspect ratio could be ten or more:

$$A_{R_1} = \frac{d_{max}}{d_{min}} \qquad \text{(Eq 6a)}$$

where A_R is the aspect ratio (dimensionless); d_{max} is the maximum diameter of a grain (μm); and d_{min} is the minimum diameter of a grain (μm).

The aspect ratio can also be defined as the reciprocal of the right-hand side of Eq 6(a), such that the ratio is normalized. In Eq 6(b), an equiaxed grain would again have an aspect ratio close to one, while an elongated grain would have an aspect ratio closer to zero:

$$A_{R_2} = \frac{d_{min}}{d_{max}} \qquad \text{(Eq 6b)}$$

Roundness relates the area of a grain to its perimeter, as in Eq 7. A perfectly circular grain would have a roundness of one, while an oblong or jagged-edged grain would have a roundness closer to zero. The roundness, like the second aspect ratio, is normalized between zero and one:

$$R = \frac{4\pi A}{P^2} \quad \text{(Eq 7)}$$

where R is roundness; A is the area of the grain (μm^2); and P is the perimeter of the grain (μm).

The grain shape index, I_s, is the mean perimeter, $\overline{L}_p$, divided by the mean intercept length, $\overline{L}_2$, as shown in Eq 8. The complexity index, I_c, in Eq 9, similar to the roundness, is affected by the jaggedness of a grain. P_L is the quantity of boundary intersections per unit length of test line, N_A is the quantity of interceptions of grains per unit test area, A_A is the areal fraction of grain-boundary interceptions, and $\overline{A}$ is the mean grain area. These and some additional shape factors are described in more detail in Ref 1:

$$I_s = \frac{\overline{L_P}}{\overline{L_2}} = \frac{\pi P_L^2}{2N_A} \quad \text{(Eq 8)}$$

$$I_c = \frac{\overline{L_P}}{\overline{A}} = \frac{\pi P_L}{2A_A} = \frac{\pi}{\overline{L_2}} \quad \text{(Eq 9)}$$

Fig. 9.4 Isopressed "opaque" (doped with MnO_2) 96% Al_2O_3 pump shaft made in Scotland. Thermally etched 3 min at 1500 °C. From the same specimen as Fig. 8.12

Example 9.5: What are the aspect ratio and roundness of the largest grain in Fig. 9.4?

Solution. The grain is approximately 23 μm long and 9 μm wide, from the scale beneath the image. The perimeter is approximately 58 μm and the area approximately 187 μm^2. One way to manually estimate the area is to photocopy the image onto a sheet of fine graph paper, count the squares in the grain, and compare the squares to the scale. Another is to photocopy the image, cut out the grain, cut out a square the length of the scale on one side, weigh the two areas, and calculate the proportion. A third is to perform a stereological point count (Ref 9). Of course, all these measurements and calculations are much more easily done by an image analyzer than manual methods. From Eq 6(b), the aspect ratio is:

$$A_R = \frac{d_{min}}{d_{max}} = \frac{9\ \mu m}{23\ \mu m} = 0.38$$

or a 2.6 to 1 ratio, which is similar to most of the grains. From Eq 7, the roundness is:

$$R = \frac{4\pi A}{P_2} = \frac{4\pi(187)}{(58)^2} = 0.699$$

Porosity and Second-Phase Content

Secondary phases, similar to grain size and shape, can affect the bulk properties of ceramics, including strength, toughness, electrical resistivity, and thermal conductivity. Second-phase content is indicative of composition and thermal process history of ceramics. Secondary phase, in this context, implies a minor constituent in a microstructure, regardless of the order of formation of phases. Some examples of secondary phases in ceramics are:

- Silicon matrix in reaction-bonded SiC
- Zirconia particles in an Al_2O_3 matrix in zirconia-toughened alumina
- Silicon carbide whiskers in an Al_2O_3 matrix composite
- Tetragonal and monoclinic grains precipitating on the grain boundaries in partially stabilized ZrO_2
- Glass between grains in 85 to 98% Al_2O_3
- Porosity in porcelain, refractories, alumina, zirconia, and so on

Second-phase content can be measured manually, as described in ASTM E 562, in which a grid is superimposed on a field of view, and intersections between grid points and secondary-phase particles are counted. The grid spacing and image magnification should be such that two grid points rarely lie on the same second-phase particle. An automatic count is possible by image analysis, as described in ASTM E 1245. The underlying assumption in stereological second-phase content measurement is that the volume fraction of a phase is equal to its point fraction in a polished cross section. Normally, porosity is measured on an unetched surface, such as that shown in Fig. 9.5(a). Secondary phases are sometimes enhanced by etching.

The mass fraction of a secondary phase is not equivalent to its volume fraction, unless the two phases have the same density. The mole fraction of a secondary phase is not equivalent to either the volume fraction or the mass fraction. Commercial ceramics are usually designated in terms of mass fraction; that is, a 96% Al_2O_3 is 96 mass% Al_2O_3 and nominally 4 mass% secondary phases, such as glass and spinel. Because SiO_2-rich glass is approximately half as dense as Al_2O_3, a 96% Al_2O_3 with 4 mass% glass would have nearly 8 vol% glass.

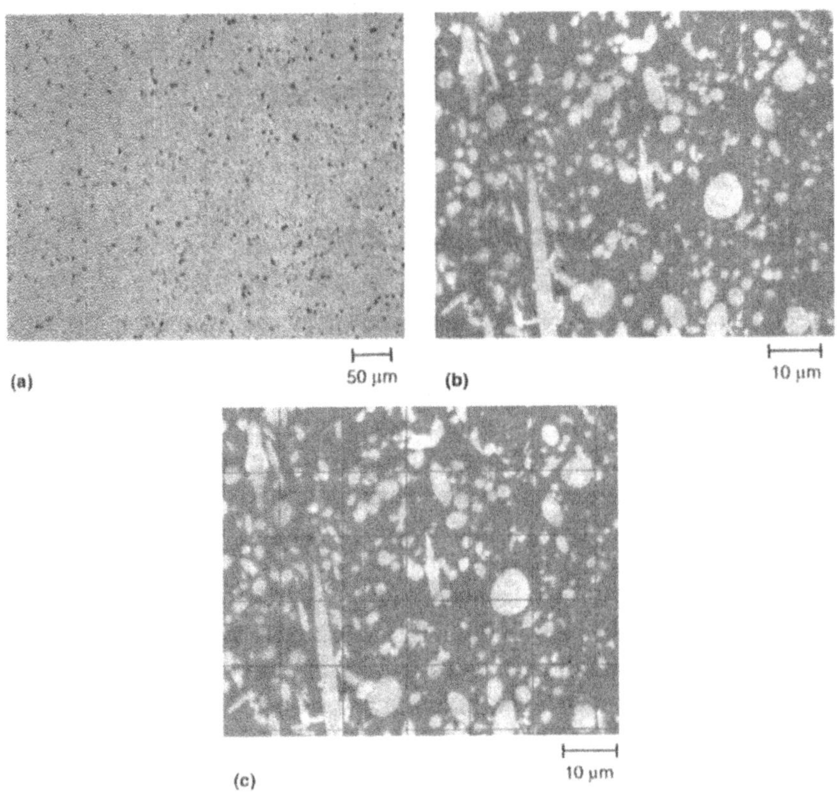

Fig. 9.5 Porosity and second-phase content. (a) Porosity in a tape-cast 96% Al_2O_3 substrate made in Brazil. Most of the pores (dark spots) are 5 μm or less in diameter. (b) Randomly oriented SiC whiskers (light phase) in an Al_2O_3 matrix. The second-phase (whisker) content is specified by the manufacturer as 30 mass%. (c) Point-counting grid superimposed on the whisker-reinforced alumina in (b)

Porosity, which can be considered as a secondary "phase," indicates the degree of densification of a ceramic. As a rule of thumb, porosity less than approximately 8% (greater than 92% densified ceramic) means that the pores are discrete, while porosity greater than 8% indicates a continuous network of pores. Porosity measurements by some other methods, such as pycnometry or buoyancy, are sensitive only to open pores and exclude occluded (closed to the surface) pores. Thus, these other porosity methods do not necessarily agree closely with measurements by ceramographic techniques. Porosity decreases strength by two important mechanisms:

- Pores reduce the true cross section area of a member.
- Pores act as stress-concentrating notches.

Porosity affects other bulk properties as well, such as the bulk modulus of elasticity and fracture energy. Porosity typically has an exponential or semilogarithmic relationship to mechanical properties, that is, $E \propto e^{-bP}$, where E is the elastic modulus or other mechanical property, b is a constant, and P is the pore volume fraction (Ref 10). The effects of pores on ceramic properties are described in detail in Rice's book on porosity.

Example 9.6: Verify the SiC content of the whisker-reinforced alumina in Fig. 9.5(b).

Solution. An evenly spaced grid is superimposed on the figure, as shown in Fig. 9.5(c). In practice, the grid would be a transparent acetate photocopy of either a sheet of graph paper or a grid drawn with the aid of drafting software overlaying a micrograph. Grid intersections that lie squarely on a second-phase particle are counted as one, such as the intersection directly above the scale bar. An intersection that lies on the interface between the primary and secondary phases is counted as 1/2, such as the top left intersection. The grid has 30 intersections, approximately 7.5 of which lie on whiskers. The volume fraction is equivalent to the area fraction, which is equivalent to the point fraction:

$$V_{\text{SiC}} = A_{\text{SiC}} = P_{\text{SiC}} = \frac{7.5}{30} = 0.25 \text{ or } 25\%$$

The mass fraction is calculated from the volume fractions and the density of each phase. The volume fraction of alumina is equal to one minus the volume fraction of whiskers, if neglible porosity is assumed:

$$M_{\text{SiC}} = \frac{(V_{\text{SiC}})(\rho_{\text{SiC}})}{(V_{\text{SiC}})(\rho_{\text{SiC}}) + (V_{\text{Al}_2\text{O}_3})(\rho_{\text{Al}_2\text{O}_3})}$$

$$M_{\text{SiC}} = \frac{(0.25)(3.17\frac{\text{g}}{\text{cm}^3})}{(0.25)(3.17\frac{\text{g}}{\text{cm}^3}) + (1 - 0.25)(3.95\frac{\text{g}}{\text{cm}^3})} = 0.211$$

This evaluation is somewhat less than the 30% specified by the manufacturer. A mere 30 grid points do not result in a very reliable measurement but could be used for a rough estimate. From Table 1 in ASTM E 562, one would have to count approximately 100 whiskers on 250 to 500 grid points on several images just to get a 10% coefficient of variation. The 30-point grid used in Fig. 9.5(c), then, would have to be applied to 10 to 15 micrographs from different regions of the specimen.

Microindentation Hardness

Microindentation hardness is a measurement of the size of a microindentation made by a diamond pyramid-shaped indenter of specified size and shape pressed into a polished surface by a known load. The surface is normally not etched prior to the indentation. The Knoop indenter has only two-fold symmetry and is commonly used on ceramics. The Vickers indenter has four-fold symmetry but makes a deeper indentation and is more inclined to cause fractures in brittle materials than the Knoop indenter.

Although *microhardness* is a widely accepted term, the more accurate name is *microindentation hardness* (Ref 11). The emphasis on microstructure and microscopic indentation size is largely what distinguishes microindentation hardness from other hardness scales, such as Rockwell and Brinell. Microindentation hardness can be used to measure the hardness of individual grains, very small pieces, and thin layers.

Hardness is not a unique property but a measure of the reaction of the ceramic to the type of disturbing force imposed (Ref 12). In addition to the indentation methods mentioned here, hardness has also been defined in terms of resistance to scratching (e.g., the Mohs scale), plowing, cutting, abrasion, erosion, damping, and rebound (e.g., Shore hardness). There is no definite numerical or even ordinal correspondence between one hardness scale and another.

Most microindentation hardness testers can perform either Knoop or Vickers hardness tests; only the indenter needs to be changed. The inden-

Table 9.1 Microindentation hardnesses of some common ceramics

Ceramic	Formula	Knoop hardness, kg/mm²
Alumina	Al_2O_3	2100
Beryllia	BeO	1250
Boron carbide	B_4C	2750
Diamond	C	7000
Silicon carbide	SiC	2480
Tantalum carbide	TaC	2000
Titanium carbide	TiC	2470
Titanium nitride	TiN	1800
Tungsten carbide	WC	1880
Tungsten carbide, cemented	WC/Co	1400–1800
Zirconia	ZrO_2	1160
Zirconium carbide	ZrC	2100

Note: The test conditions are unspecified. Source: Ref 14

ter changeover may require some realignment of the instrument and should be done as infrequently as possible. The indenter is a precisely machined, microscopic diamond attached to a steel mandrel. The indenter should be handled and stored very carefully to avoid damage. A worn-out or damaged indenter cannot be repaired; it must be replaced. Misshapen indentations may be an indication of a worn-out or damaged indenter. An asymmetric indentation is usually an indication of a tilted surface.

The microindentation hardness tester is very sensitive to vibrations and should be acoustically isolated in the laboratory. A granite table, the kind used for precision electrobalances, is an ideal site for a microindentation hardness tester. An air suspension table is also suitable. The microindentation hardness tester is an optical instrument and, similar to the microscopes, should be located in a clean room apart from the preparation equipment and any other dust-generating machinery.

The microindentation hardness of a coating can be measured directly if the coating is smooth and flat enough for microscopic examination and thick enough such that its substrate does not interfere. Another, and often better, way to measure the microindentation hardness of a coating is by cross section, which is independent of the flatness and smoothness of the coating and interference by the substrate. Thin ceramic plates, such as those used for microelectronic packaging, are also best measured for hardness in cross section, in most cases.

Most microindentation hardness measurements require a polished section; polishing methods are described in Chapters 3 and 4. The microindentation hardness should be measured prior to etching, relief polish, or removal of the specimen from its encapsulation. Edge retention in the mount is very important when the microindentation hardness of a surface layer or other edge region is to be measured.

The load on the indenter can have an effect on the hardness measurement, especially at low loads (Ref 13). Low loads, say less than 100 g on ceramics, tend to result in very high microindentation hardness values. The hardness values in Table 9.1 were probably measured under a very small load, whereas the values in Tables 9.2 and 9.3 were measured at much higher loads. Furthermore, the small indentation made by a low load can be difficult to measure precisely. Very high loads tend to cause cracks in the corners of the indentation, which also affect the observed hardness.

An inverse square root relationship between hardness and grain size, similar to Eq 2, has been observed in a number of ceramics over a range

Table 9.2 Microindentation hardnesses of selected common ceramics

Ceramic	Formula	Indenter load, g	Dwell time, s	Knoop hardness, kg/mm²
Sialon	$Si_3N_4 \cdot Al_2O_{3(ss)}$	1000	10	1350
Silicon carbide, reinforced	SiC + 1.5% AlN	1000	10	2017
Strontium-celsian (Ref 15)	$SrAl_2Si_2O_8$ monoclinic	300	10	466
Zircon	$ZrSiO_4$	500	15	238

Table 9.3 Microindentation hardness, toughness, elastic (Young's) modulus, and density of some common ceramics

Ceramic	Formula	Knoop hardness (*H*)(a) kg$_f$/mm^2	Knoop hardness (*H*)(a) GPa	Toughness (K_c), MPa$\sqrt{m}$(b)	Elastic modulus (*E*), GPa(c)	Density (ρ), g/cm^3
Alumina, 85%	α-Al_2O_3	960	9	3–4	221	3.41
Alumina, 90%	α-Al_2O_3	1058	10	3–4	276	3.60
Alumina, 94%	α-Al_2O_3	1175	12	3–4	296	3.70
Alumina, 96%	α-Al_2O_3	1088	11	3–4	303	3.72
Alumina, 99.5%	α-Al_2O_3	1440	14	3–4	372	3.89
Alumina, 99.9%	α-Al_2O_3	1551	15	3–4	386	3.96
Alumina, reinforced with SiC whiskers and hot pressed	α-Al_2O_3 + SiC_w	2075	20	8	424	3.80
Alumina, reinforced with SiC whiskers and pressureless-sintered	α-Al_2O_3 + SiC_w	2050	20	5	344	3.80
Boron carbide	B_4C	3000	29	3–4	350	2.50
Mullite	$3Al_2O_3 \cdot 2SiO_2$	750	7	2	150	2.80
Silicon carbide, direct sintered	SiC	2800	27	4	400	3.10
Silicon carbide, reaction bonded	SiC	2500	25	3–4	393	3.10
Silicon nitride, hot pressed	Si_3N_4	1500	15	6	311	3.31
Silicon nitride, reaction bonded	Si_3N_4	1350	13	3–4	179	2.5
Spinel, optical grade	$MgAl_2O_4$	1300	13	...	260	3.58
Titanium diboride	TiB_2	2700	26	5	538	4.48
Tungsten carbide	WC	1600–2050	16–20	3–28	483–607	14–15
Zirconia, TTZ	ZrO_2 (toughened)	1200	12	11	200	5.75
Zirconia, Y-TZP	$ZrO_2 + Y_2O_3$ (tetragonal)	1300	13	13	200	6.02
Zirconia, ZDY	$ZrO_2 + Y_2O_3$ (cubic)	...	...	3	173	5.60

(a) Determined via ASTM E 384 under a 1000 g load. The Knoop hardness in GPa is approximately equivalent to the diamond pyramid hardness in Eq 14 and 15. (b) Determined by a notched-beam method, which is not necessarily equivalent to the indentation methods. (c) Determined via ASTM C 848 at 20 °C. Source: Ref 16

of grain sizes and indentation loads (Ref 17). That is, the ceramics with the smallest grain sizes were the hardest. A hardness minimum was noted in most ceramics when the indentation is approximately the same size as the grains, due to spalling and cracking along the grain boundaries.

ASTM procedures that apply to microindentation hardness, including C 849, C 1326, E 6, E 92, and E 384, are listed in Appendix A. Reference books on microindentation hardness include one from McColm and several published by ASM International.

Knoop hardness uses a pyramid with two 136° base angles and loads up to approximately 1000 g, as in Fig. 9.6(a). The microindentation hardness is calculated from the length of the longer axis only of the indentation, as shown in Fig. 9.5(b) and (c), and the applied load. The hardness is expressed as a Knoop hardness number (HK or HKN) in Eq 10 or in the Système International d'Unités (SI) pressure unit, gigapascal, in Eq 11:

$$\mathrm{HK}\left[\frac{\mathrm{kg_f}}{\mathrm{mm}^2}\right] = 14229\ \frac{P}{d_2} \qquad \text{(Eq 10)}$$

$$HK(GPa) = 0.0098067 \times HK\left[\frac{kg_f}{mm^2}\right] = 139.54\,\frac{P}{d_2} \qquad \text{(Eq 11)}$$

where HK is the Knoop hardness number, in the bracketed units; P is the indenter load (g) and d is the length of the longer diagonal of the Knoop indentation (μm).

The constants in Eq 10 and 11 take into account the area projected by the indenter and the unit conversion factors. The error in the diagonal length measurement has been liberally estimated, perhaps overestimated, from the wavelength, λ, and the numerical aperture, NA, as $\Delta d = 7\lambda/(2NA)$ (Ref 13). Some Knoop hardnesses for ceramics are given in Tables 9.1, 9.2, and 9.3. A procedure for Knoop hardness measurement is shown as follows:

1. Prepare a polished section of the ceramic to be measured, as in Chapters 3 to 4. Do not use relief polish, do not remove the specimen from the encapsulation, and do not etch or sputter coat the specimen. Compression-mounted specimens are preferable to castable resins, because the opposing faces are flat and parallel. The unpolished face of a cast mount has a meniscus (as in Fig. 3.6) that precludes flatness without additional machining or a layer of acrylic applied in compression.
2. Verify the accuracy of the instrument by indenting a test block according to the manufacturer's recommendations. If the measured hardness of the test block is outside the acceptable range, the instrument should be professionally tuned. If the indentation is misshapen, the indenter should be replaced. Perform the calibration check at least once each day that the hardness tester is used.
3. Clamp the mounted specimen in the vise of the hardness instrument. The polished surface must be level and rigidly supported. Most hardness testers have a self-leveling vise designed to accommodate 25 and 30 mm mounted specimens. Do not use modeling clay beneath the specimen, as in Subroutine 7.1, or anything else that will yield under the indenter load.
4. Set the load and dwell time on the hardness tester. For most fully densified ceramics, a 500 or 1000 g load and 10 or 15 s dwell time are recommended. Higher loads increase the likelihood of cracked indentation corners and spalling. Lower loads make smaller indentations that can be difficult to measure precisely. The dwell time must be brief enough to minimize the effects of vibrations on the indenter yet long enough to minimize anelastic effects.
5. Focus on the site on the polished surface where the indentation is to be made. Adjust the optics for Köhler illumination, if possible. A green filter in the light path minimizes eye fatigue. Move the stage or turret as appropriate to the indentation position and actuate the indenter. When the indentation is complete, return the turret or stage to the viewing position.

6. Align the filar eyepiece with the axes of the indentation and measure the long axis. Record the reading. Disregard an indentation with badly cracked corners, a highly asymmetric indentation, or an indentation that intersects a large pore.
7. Adjust the stage position for a new indentation and repeat steps 5 and 6. A minimum of thirty indentations is necessary for a normal distribution, but ten is adequate for most purposes. A coefficient of variation of 10 to 15% for hardness measurements on one ceramic specimen is not unusual. Some typical Knoop hardnesses of ceramics are given in Tables 9.1, 9.2, and 9.3.

Example 9.7: What is the Knoop hardness of the ceramic in Fig. 9.6(b)?

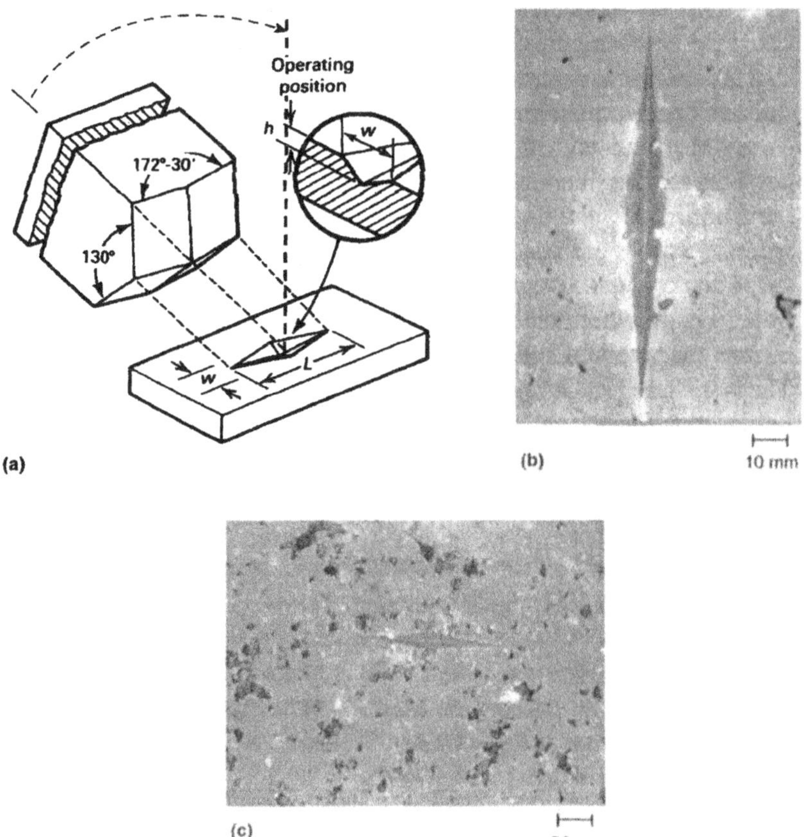

Fig. 9.6 Knoop hardness testing. (a) Schematic of a Knoop hardness indenter. The depth *h* of the indentation is approximately 1/30 of the length *L* of the longer diagonal (Ref 11). (b) Knoop indentation in 99.5% Al_2O_3 1000 g load. (c) Knoop indentation in 96% "opaque" (tinted brown by MnO_2) Al_2O_3 500 g load. The indentation is approximately 82 μm long; therefore HKN_{500} = 1058 kg_f/mm^2.

Solution. The applied load was given as 1000 g. The longer of the two axes is approximately 102 μm, according to the scale marker. From Eq 10:

$$HK = 14229 \frac{1000 \text{ g}}{(102 \text{ μm})^2} = 1368 \frac{\text{kg}_\text{f}}{\text{mm}^2}$$

and from Eq 11:

$$HK = 139.54 \frac{1000 \text{ g}}{(102 \text{ μm})^2} = 13.4 \text{ GPa}$$

Some older machines require the operator to make these calculations or to read them from a table, while most newer machines make the calculation automatically and report the Knoop hardness number. The Knoop hardness number conversions are given in Table 1 of ASTM E 384.

Vickers hardness uses a pyramid with 90° base angles and 136° face angles (α), as in Fig. 9.7(a), and loads up to approximately 1000 g. Vickers is not recommended for microindentation hardness measurements of most sintered ceramics, because the indentations are very small and tend to crack the surrounding regions such that the true size of the indentation is difficult to determine accurately. The hardness is calculated from the mean length of the two diagonals, as shown in Fig. 9.7(b) and (c), and the applied load. The calculation is given in ASTM E 384 as Eq 12(a), which is modified here as Eq 12(b) and (c). The constant in Eq 12(a) accounts for the contact area of the indenter (2 sin(α/2)) and the unit conversion factors. The constants in Eq 12(b) and (c) also include a factor of 4 to account for the average of the two diagonals. The procedure is essentially the same as the Knoop procedure, except that both axes of the indentation are measured:

$$HV \left[\frac{\text{kg}_\text{f}}{\text{mm}^2} \right] = \frac{2P_{kg} \sin(\alpha/2)}{(\bar{d}_{mm})^2} = 1.8544 \frac{P_{kg}}{(\bar{d}_{mm})^2} \quad \text{(Eq 12a)}$$

where the load P is in kilograms, the mean of the two diagonals $\bar{d}$ is in millimeters, and the angle between opposite faces $\alpha = 136°$:

$$HV \left[\frac{\text{kg}_\text{f}}{\text{mm}^2} \right] = 7417.6 \frac{P}{(d_1 + d_2)^2} \quad \text{(Eq 12b)}$$

$$HV(\text{GPa}) = 0.0098067 \times HV \left[\frac{\text{kg}_\text{f}}{\text{mm}^2} \right] = 72.742 \frac{P}{(d_1 + d_2)^2} \quad \text{(Eq 12c)}$$

where HV is the Vickers hardness number, in the bracketed units; P is the indenter load (g); d_1 is the length of the first diagonal of the Vickers indentation (µm); and d_2 is the length of the second diagonal of the Vickers indentation (µm).

Diamond Pyramid Hardness. A third type of microindentation hardness is diamond pyramid hardness, which is the same as Vickers except that the applied load ranges from approximately 1 to 50 kg. Diamond pyramid hardness is the basis for fracture toughness measurement by microindentation, discussed in a subsequent section. The diamond pyramid hardness is defined in Eq 13 and calculated from Eq 12(b) and (c):

$$\text{DPH (kg}_f\text{/mm}^2 \text{ or GPa)} = \text{HV (kg}_f\text{/mm}^2 \text{ or GPa)} \qquad \text{(Eq 13)}$$

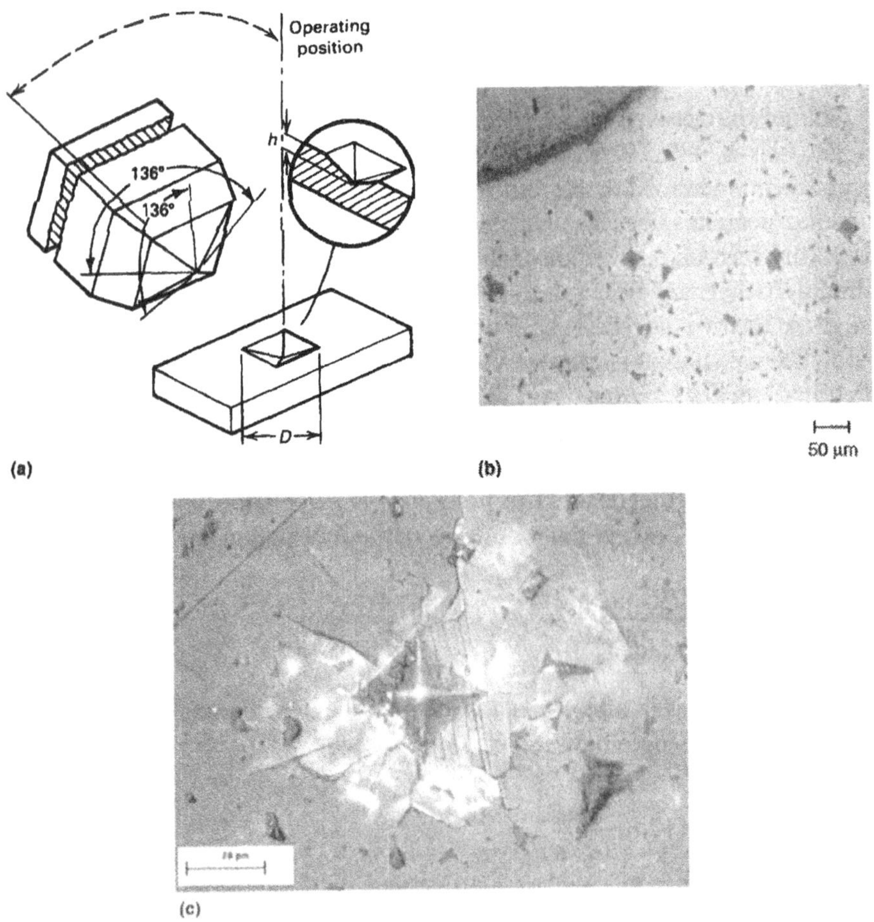

Fig. 9.7 Vickers hardness testing. (a) Schematic of a Vickers or diamond pyramid hardness indenter. The indentation depth *h* is approximately 1/7 of the length of the average length *D* of the diagonals (Ref 11). (b) Vickers indentations on a radial cross section of an extruded 99.8% Al_2O_3 thermocouple sheath that had crept after numerous furnace cycles. The load on the indenter was 1.000 kg, and the dwell time was 15 s. (c) Magnified view of the left Vickers indentation in (b). Note the fracture field surrounding the indentation and the lower contrast at this magnification compared to the lower magnification. Photo by P.E. Danielson

where DPH is the diamond pyramid hardness number and HV is the Vickers hardness number, in the same units.

Example 9.8: What is the Vickers hardness of the ceramic shown in Fig. 9.7(c)?

Solution. The applied load was given as 1000 g. The vertical axis is approximately 36 μm, according to the scale marker. The horizontal axis is approximately 34 μm. From Eq 12(b):

$$HV = 7417.6 \frac{1000\ \text{g}}{(36\ \mu\text{m} + 34\ \mu\text{m})^2} = 1514 \frac{\text{kg}_\text{f}}{\text{mm}^2}$$

and from Eq 13:

$$HV = 7417.6 \frac{1000\ \text{g}}{(36\ \mu\text{m} + 34\ \mu\text{m})^2} = 14.8\ \text{GPa}$$

Some older machines require the operator to make these calculations or to read them from a table, while most newer machines make the calculation automatically and report the Vickers hardness number. The Vickers hardness number conversions are given in Table 2 of ASTM E 384.

Rockwell Superficial Hardness. Rockwell hardness is a macroindentation rather than microindentation hardness scale and requires a smooth but not necessarily polished surface. The hardness value is a function of the difference in indentation depth between a minor applied load and a major one. The Rockwell superficial scales, particularly the 45N scale, are applicable to ceramics and other brittle materials. The hardest metals, such as martensitic steel, are usually measured with the Rockwell C scale, in which the major load is 150 kg. Most structural ceramics are much harder than any steel but tend to fracture under the severe load of the Rockwell C diamond indenter. In the Rockwell 45N scale, a 45 kg load is applied to the cone-shaped diamond indenter. The as-fired surface of high-alumina devices is usually smooth enough for Rockwell superficial hardness. A curved surface, such as the outside diameter of an alumina or mullite rod, can be tested with the Rockwell 45N scale. The dimensionless Rockwell hardness value is read directly from the gage of the indentation machine. There are no conversion factors that directly relate Rockwell hardness to any of the microindentation hardness scales. As with the Knoop procedure mentioned previously, a minimum of ten indentations per specimen is recommended for a hardness report.

The procedure for Rockwell superficial hardness is given in ASTM E 18, listed in Appendix A.

Toughness

Toughness is the ability of a material to resist propagation of fractures. Softer and more ductile materials, such as most metals and elastomers, are generally tougher than harder and more brittle materials, including most ceramics. Toughness can be measured with impact testers and strength testers, but ceramographic methods have been devised that use a diamond pyramid indentation under a load of 4 to 20 kg.

In the indentation toughness methods, ceramic bars are prepared for modulus-of-rupture (MOR) testing, and the tensile face of each bar is polished to ceramographic standards. The polished tensile face is diamond pyramid indented in three places, evenly spaced, along the centerline of the bar. The bar is secured in the test fixture in a strength-testing machine and loaded continuously to the point of fracture. Cracks grow perpendicular to the tensile direction from the corners of all three indentations while the bar is being loaded by the strength-testing machine. The indentations act as artificial critical flaws, and the fracture path normally includes one of the indentations. The toughness is calculated from the dimensions of the bar, the indentation load, the lengths of the cracks emanating from the indentations, the applied force at the instant of failure, and the hardness and elastic modulus of the ceramic. The correlation between actual toughness and the parameters measured in the indentation method is not exact, which is why more than one formula has been published. Three calculations are presented here.

Modulus of rupture is a bending strength test widely used in the ceramics industry, because pure tension is difficult to apply to brittle materials. In MOR, the movable head of a strength-testing machine presses one or two closely spaced hardened steel rollers into a bar supported between two widely spaced steel rollers on a fixed platform. If one top roller is used, it is called three-point bending, because the stresses are applied to the bar by contact with three rollers. If two top rollers are used, it is four-point bending, shown schematically in Fig. 9.8(a). The four-point fixture is preferred, because it puts a larger volume of the bar in tension and yields more reliable results. The bottom of the bar, in contact with the two widely spaced rollers, is in tension. The top of the bar, in contact with two closely spaced rollers or only one roller in the center, is in compression. The strength-testing machine measures the force it is applying to the bar with a load cell. The deflection of the bar can be measured with a strain gage or extensometer, although it is not necessary for the indentation toughness test.

Transformation-toughened (TTZ) or partially stabilized zirconia (PSZ) uses a phase-change mechanism triggered by impact energy to achieve unusually high toughness for a ceramic. Zirconia with CaO or MgO in solid solution retains the metastable cubic structure on cooling, but tetragonal grains nucleate while cooling through the tetragonal phase field. The tetragonal precipitates, which are normally stable from approximately 1170 to 2370 °C, can be stabilized down to room temperature by heat treatments or other means. The tensile stress field ahead of a crack tip transforms the metastable, ~0.3 μm tetragonal grains into the stable monoclinic phase in a martensitic reaction. The less-dense monoclinic phase, constrained by the cubic matrix, tries to expand but cannot. The constraint puts the crack tip in compression so that it cannot extend (Ref 18, 19). This phase-transformation mechanism makes PSZ an unusually tough ceramic that can be used for impact applications.

Many factors from both the microstructure and the test conditions can affect the toughness measurement. The grain size, grain shape, porosity, anisotropy, and other microstructural characteristics can be expected to influence the measured toughness. The indenter load, crack size, crack orientation, MOR test fixture, crosshead speed, surface finish, and other test conditions are likely to have an effect on the toughness results. The cracks often grow in a tortuous path and can be difficult to measure.

Toughness measured by the Anstis et al. (Ref 20) crack-length method is dependent on the elastic modulus of the material, microindentation hardness and crack length, and the applied load, as in Eq 14. Cracks

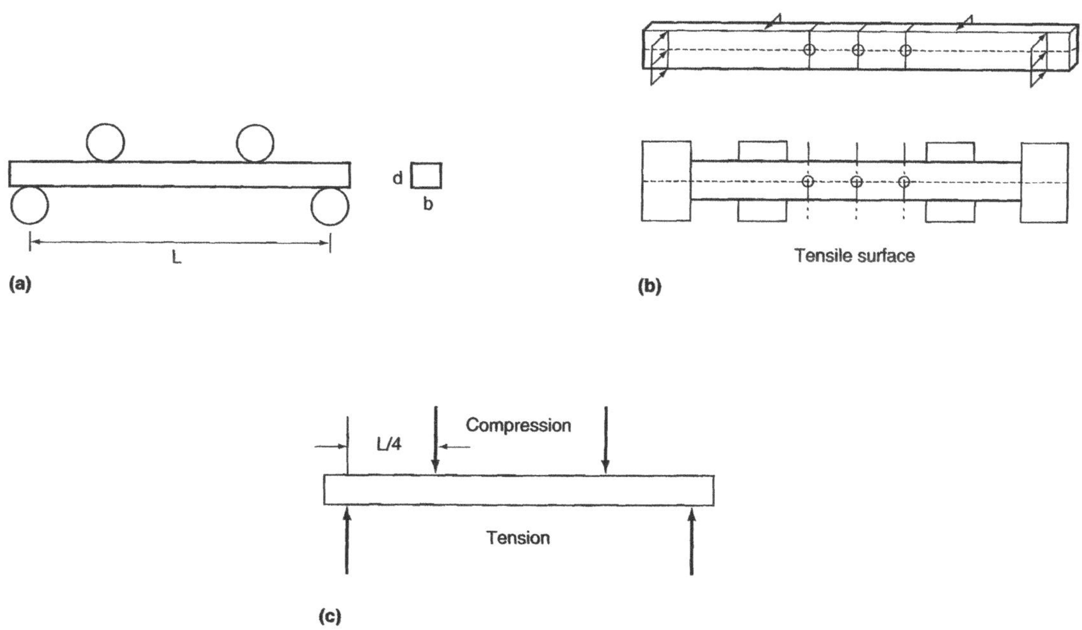

Fig. 9.8 Modulus of rupture (MOR) toughness testing. (a) Four-point MOR bending fixture. In configuration B, $L = 40$ mm, $b = 4$ mm, and $d = 3$ mm. The upper pins are 20 mm apart on center. (b) Three diamond pyramid hardness indentations in the tensile surface of a MOR bar for toughness measurement. The indentations are symmetric in the x- and y-directions. The indentation diagonals are parallel or perpendicular to the tensile direction. (c) Pin positions in a four-point MOR bending fixture. $L/4 = 10$ mm in a configuration-B bar.

emanate from the corners of the indentation and form a semicircle underneath the indentation in the *y*-*z* and *z*-*x* planes:

$$K_c = 0.016\sqrt{\frac{E}{H}\,\frac{P}{(c_o)^{3/2}}} \qquad \text{(Eq 14)}$$

where K is toughness (MPa$\sqrt{m}$ or MN/m$^{1.5}$) (the subscript indicates the calculation method); E is elastic modulus (GPa); and H is diamond pyramid hardness (GPa); and P is the indentation load (MN). Multiply the load in kilograms by 9.8067×10^{-6} to get meganewtons. c_0 is half the average length of the initial cracks (m). Multiply the sum of the two crack lengths in microns by 2.5×10^{-7} to get half the mean crack length in meters.

The Chantikul et al. (Ref 21) method uses the tensile strength, or indentation strength in bending (isb), instead of the crack length, as shown in Eq 15, where σ is the applied tensile stress (MPa) at fracture:

$$K_{isb} = 0.59\left(\frac{E}{H}\right)^{1/8}[\sigma(P^{1/3})]^{3/4} \qquad \text{(Eq 15)}$$

The Cook and Lawn (Ref 22) method uses the maximum applied stress and the extended crack length (ecl) of the two surviving indentations after the bars are broken, as shown in Eq 16. Normally, the bars break through one of the three indentations, and the cracks that are perpendicular to the tensile stress grow as the flexural stress is applied. If the fracture path does not include one of the indentations, it is an indication that the bar contains fabrication flaws in the tensile zone that are larger or sharper than the indentations:

$$K_{ecl} = 2.02\sigma_m\sqrt{c_m} - 0.68 \qquad \text{(Eq 16)}$$

where σ_m is the maximum applied stress (MPa); and c_m is half the average crack length (m) of the cracks that extend perpendicular to the direction of the tensile stress, emanating from the corners of the two surviving indentations.

A procedure for toughness measurement by these methods is given as follows:

1. Prepare a minimum of ten MOR bars, nominally 45 mm long with a rectangular cross section 3 by 4 mm. Specifications are given in ASTM C 1161, configuration B; Military Standard (MIL-STD) 1942 (MR); the German specification Deutsches Institut für Normung (DIN) 51-110-1; and the French B41-104. These bars are normally used for flexural strength testing in a four-point bending fixture, as shown in Fig. 9.8(a). Other bar sizes are also permitted in these specifications. Three-point bending fixtures can also be used, but four-

point bending puts a larger volume of the specimen in tension, resulting in a lower standard deviation and more reliable data. The larger ceramic fabricators and many specialty contractors have machine shops that can precisely saw, grind, and polish MOR bars.

2. Polish one 4 by 45 mm face with successively finer abrasives to a roughness of approximately 4 μm R_a or better, as measured by a profilometer. Normally, 1 μm diamond is the finest abrasive needed.
3. Indent the polished face with a diamond pyramid indenter under a 20 kg load in three places along a 45 mm centerline, uniformly spaced between the innermost pins of the bending fixture, as shown in Fig. 9.8(b). Other loads can be used, with the risk of not enough damage by a lesser load or too much damage by a greater load. The P term in Eq 14 becomes 1.962×10^{-4} MN when the indenter load is 20 kg.
4. A dye penetrant makes the cracks more visible. Draw a stripe across the indentations with the broad edge of a felt-tip marking pen, then quickly wipe the excess ink off with a Kim-Wipe (Kimberly-Clark Corp.) or similar lab tissue. Measure all three pairs of crack lengths with a graduated reticle in a microscope or with the filar eyepiece of the hardness tester and record the results.
5. Break each bar with the polished side down, thus in tension, in a strength-testing machine equipped with a four-point bending fixture. The fixture should have spans of 40 mm between the centers of the outer, lower pins and 20 mm between the centers of the upper, inner pins, as shown in Fig. 9.8(c). Save the load-versus-deflection curve and record the load at failure. Save and number each broken bar for further analysis.
6. Measure the width and thickness of each bar next to the fracture. Calculate the indentation strength in (four-point) bending from Eq 17:

$$\sigma = \frac{3P \cdot L}{4b \cdot d^2} \qquad \text{(Eq 17)}$$

where: σ is the indentation strength in bending or modulus of rupture (MPa); P is the load at fracture (N), measured by the load cell of the strength-testing machine; L is the length (mm) between the outermost pins, nominally 40 mm; b is the bar width (mm), nominally 4 mm; and d is the bar thickness (mm), nominally 3 mm.

For a configuration B bar, Eq 17 is approximated by $\sigma \approx 0.8333 \times P$.

7. Measure again the crack lengths of the two surviving indentations, as in step 4, for Eq 16. The fracture surface of a failed indentation is shown in Fig. 9.9(a). The extended cracks of surviving indentations from three ceramics are shown in Fig. 9.9(b), 9.10(a), and 9.10(b).
8. Calculate the toughness from either Eq 14, 15, or 16. These three equations do not necessarily yield the same result and typically do not

agree exactly with toughness results from other methods, such as three-point bending or single-edge notched beam. The stress (σ) term in Eq 15 and 16 comes from Eq 17. The E and H terms in Eq 14 and 15 are usually taken from the literature, such as Table 9.3. The diamond pyramid hardness (H) can also be measured, as described in a previous section.

Example 9.9: What is the toughness of the material around the indentation in Fig. 9.10(a)?

Solution. From Table 9.3, $E = 344$ GPa and $H = 20$ GPa. The horizontal crack is approximately 345 μm long, and the vertical crack is approximately 320 μm long, from the scale in the micrograph. Note that the very tips of the cracks may be too small to resolve, and that cracks sometimes terminate at pores. The indentation load conversion is:

$$P = (10\ \text{kg})\left(9.81\frac{\text{m}}{\text{s}^2}\right)\left(10^{-6}\frac{\text{MN}}{\text{N}}\right) = 9.81 \times 10^{-5}\ \text{MN}$$

Half the initial mean crack length is calculated from the mean of the two cracks:

$$c_o \left(\frac{1}{2}\right)\left(\frac{345\ \mu\text{m} + 320\ \mu\text{m}}{2}\right)\left(\frac{10^{-6}\ \text{m}}{\mu\text{m}}\right) = 1.66 \times 10^{-4}\ \text{m}$$

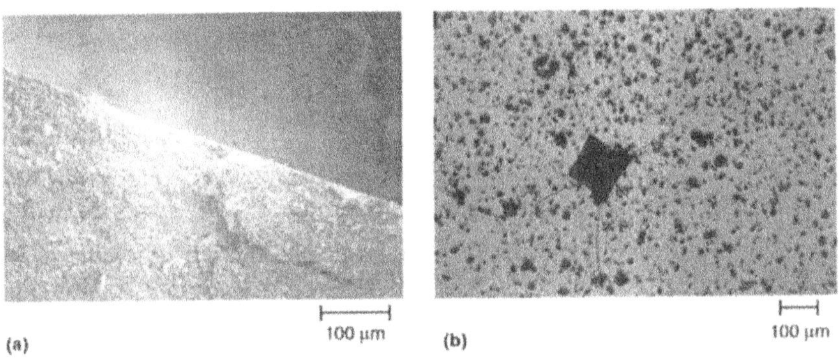

Fig. 9.9 Fracture toughness determination for 96% Al_2O_3 ceramic composite. (a) Fracture surface in an experimental 96% Al_2O_3, 4% $SrAl_2Si_2O_8$ composite (Ref 5). Indentation load = 20 kg. The indentation is the bright, broad "V" along the top edge. A half-penny crack extension is visible below and to the right of the indentation. (b) Fracture toughness indentation in an experimental 96% Al_2O_3, 4% $SrAl_2Si_2O_8$ composite. Indentation load = 20 kg. The configuration B bar fractured under a load of 172 N in four-point bending (Ref 23). The tensile stress direction is parallel to the horizontal diagonal of the indentation.

From Eq 14:

$$K_c = 0.016 \sqrt{\frac{344\ \text{GPa}}{20\ \text{GPa}}} \left(\frac{9.81 \times 10^{-5}\ \text{MN}}{(1.66 \times 10^{-4}\ \text{m})^{3/2}} \right) = 3.04\ \frac{\text{MN}}{\text{m}^{1.5}} = 3.0\ \text{MPa}\ \sqrt{\text{m}}$$

This toughness value is somewhat less than the published value in Table 9.3, which was determined by a different method. Equations 15 and 16 cannot be used here, because the indentation strength is not given and is usually much less than the published MOR value.

If the DPH in the photo was measured, the horizontal diagonal of the indentation is approximately 105 μm long and the vertical approximately 100 μm long. From Eq 12(c) and 13, DPH = 15.7 GPa. The toughness (K_c) would be slightly higher at 3.4 MPa$\sqrt{\text{m}}$. Note that the DPH is affected by the corner cracking caused by the high indentation load, which is why the literature value measured under a smaller load with little or no corner cracking is preferred.

Example 9.10: What is the toughness of the material around the indentation in Fig. 9.9(b)?

Solution. From Table 9.3, E = 303 GPa and H = 11 GPa. The vertical crack is approximately 713 μm long, from the scale in the micrograph. The indentation load conversion is:

$$P = (20\ \text{kg})\left(9.81\ \frac{\text{m}}{\text{s}^2}\right)\left(10^{-6}\ \frac{\text{MN}}{\text{M}}\right) = 1.96 \times 10^{-4}\ \text{MN}$$

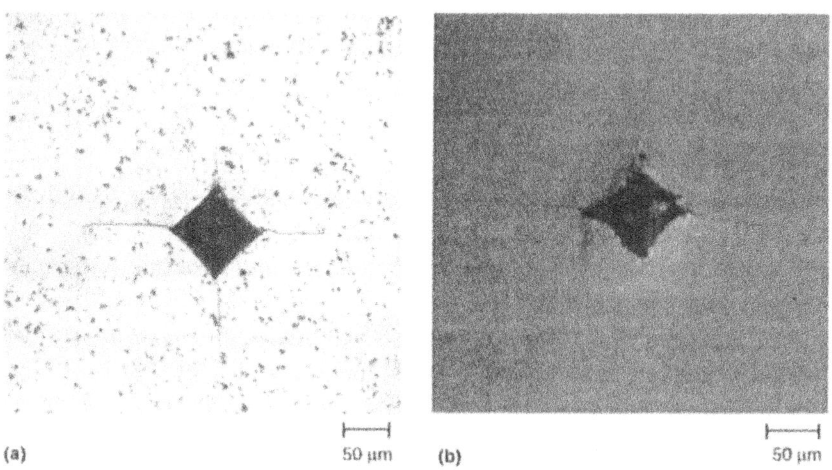

Fig. 9.10 Fracture toughness determination for SiC ceramics. (a) Fracture toughness indentation in SiC whisker-reinforced Al_2O_3 10 kg load. (b) Fracture toughness indentation in SiC + 1.5% AlN, 10 kg load. Photographed in differential interference contrast. Note the irregular shape of the indentation.

Half the initial mean crack length is:

$$c_o = \left(\frac{713\ \mu\text{m}}{2}\right)\left(10^{-6}\frac{\text{m}}{\mu\text{m}}\right) = 3.57 \times 10^{-4}\ \text{m}$$

The fracture stress, from Eq 17, is:

$$\sigma = \frac{3P \cdot L}{4b \cdot d^2} = \frac{3(172\ \text{N})(40\ \text{mm})}{4(4\ \text{mm})(3\ \text{mm})^2} = 143\ \frac{\text{N}}{\text{mm}^2} = 143\ \text{MPa}$$

Notice that the strength after indentation is less than half the mean strength of a set of bars that were not indented, 332 MPa. The indentation strength in bending method, Eq 15, yields:

$$K_{\text{isb}} = 0.59\left(\frac{303\ \text{GPa}}{11\ \text{GPa}}\right)^{1/8}\left[(143\ \text{MPa})(1.96 \times 10^{-4}\ \text{MN})^{1/3}\right]^{3/4} =$$
$$4.4\ \text{MPa}\sqrt{\text{m}}$$

The extended crack length method, Eq 16, yields a slightly higher toughness:

$$K_{\text{ecl}} = 2.02(143\ \text{MPa})\sqrt{3.57 \times 10^{-4}\ \text{m}} - 0.68 = 4.8\ \text{MPa}\sqrt{\text{m}}$$

These and other toughness methods are described and compared in Ref 24. Section 11 of *Mechanical Testing and Evaluation,* Volume 8 of *ASM Handbook,* contains hardness, toughness, and modulus values of other ceramics from a number of sources, along with other engineering properties. Numerous indentation toughness models are summarized in Ref 12.

REFERENCES

1. E.E. Underwood, *Quantitative Stereology,* Addison-Wesley Publishing Co., 1970, p 4, 23–30, 81, 227–232
2. G.F. Vander Voort, *Metallography: Principles and Practice,* American Society for Metals, 1984, p 490
3. W.D. Kingery, H.K. Bowen, and D.R. Uhlmann, *Introduction to Ceramics,* 2nd ed., John Wiley & Sons, Inc., 1976, p 739, 794
4. G.E. Dieter, *Mechanical Metallurgy,* 3rd ed., McGraw-Hill, Inc., 1986, p 189–191
5. R.W. Rice, *Mechanical Properties of Ceramics and Composites,* Marcel Dekker, Inc., 2000, p 127–244
6. M.I. Mendelson, Average Grain Size in Polycrystalline Ceramics, *J. Am. Ceram. Soc.,* Vol 52 (No. 8), 1969, p 443–446

7. J.C. Wurst and J.A. Nelson, Lineal Intercept Technique for Measuring Grain Size in Two-Phase Polycrystalline Ceramics, *J. Am. Ceram. Soc.*, Vol 62 (No. 2), Feb 1972, p 109
8. L.H. Van Vlack, *Physical Ceramics for Engineers,* Addison-Wesley Publishing Co., Inc., 1964, p 99–100
9. J.C. Russ and R.T. Dehoff, *Practical Stereology,* 2nd ed., Kluwer Academic, 2000, p 45–77
10. R.K. MacCrone, Ed., *Treatise on Materials Science and Technology,* Vol 11, *Properties and Microstructure,* Academic Press, 1977, p 199– 381
11. G.F. Vander Voort, Microindentation Hardness Testing, *Mechanical Testing and Evaluation,* Vol 8, *ASM Handbook,* ASM International, 2000, p 221–231
12. I.J. McColm, *Ceramic Hardness,* Plenum Press, 1990, p 1–11, 145–187
13. G.D. Quinn, Indentation Hardness Testing of Ceramics, *Mechanical Testing and Evaluation,* Vol 8, *ASM Handbook,* ASM International, 2000, p 244–251
14. R.C. Weast, Ed. *CRC Handbook of Chemistry and Physics,* 57th ed., CRC Press, Inc., 1976, p F-24
15. R.E. Chinn, "Transient Glass-Phase Processing of $SrAl_2Si_2O_8$-Alumina Ceramics," Thesis T-4685, Colorado School of Mines, 1995, p 88, 148
16. "Applications Guide," Coors Ceramics Company Bulletin #980, circa 1994
17. R.W. Rice, C.C. Wu, and F. Borchelt, Hardness—Grain-Size Relations in Ceramics, *J. Am. Ceram. Soc.*, Vol 77 (No. 10), 1994, p 2539–2553
18. R. Stevens, Engineering Properties of Zirconia, *Ceramics and Glasses,* Vol 4, *Engineered Materials Handbook,* ASM International, 1991, p 775–786
19. Y.-M. Chiang, D. Birnie III, and W.D. Kingery, *Physical Ceramics: Principles for Ceramic Science and Engineering,* John Wiley & Sons, Inc., 1997, p 488–492
20. G.R. Anstis, P. Chantikul, B.R. Lawn, and D.B. Marshall, A Critical Evaluation of Indentation Techniques for Measuring Fracture Toughness, Part I: Direct Crack Measurements, *J. Am. Ceram. Soc.,* Vol 64 (No. 9), Sept 1981, p 533–538
21. P. Chantikul, G.R. Anstis, B.R. Lawn, and D.B. Marshall, A Critical Evaluation of Indentation Techniques for Measuring Fracture Toughness, Part II: Strength Method, *J. Am. Ceram. Soc.,* Vol 64 (No. 9), Sept 1981, p 539–543
22. R.F. Cook and B.R. Lawn, A Modified Indentation Toughness Technique, *Comm. Am. Ceram. Soc.,* Nov 1983, p C-200–C-201
23. R.E. Chinn, M.J. Haun, C.Y. Kim, and D.B. Price, Microstructures and Properties of Three Composites of Alumina, Mullite and Monoclinic $SrAl_2Si_2O_8$, *J. Am. Ceram. Soc.,* Vol 83 (No. 11), 2000, p 2668– 2672

24. K.E. Amin, Toughness, Hardness, and Wear, *Ceramics and Glasses,* Vol 4, *Engineered Materials Handbook,* ASM International, 1991, p 599–609

SUGGESTED READING

- H.E. Boyer, Ed., *Hardness Testing,* ASM International, 1987
- *Ceramics and Glasses,* Vol 4, *Engineered Materials Handbook,* ASM International, 1991
- H. Chandler, Ed., *Hardness Testing,* 2nd ed., ASM International, 1999
- H. Kuhn and D. Medlin, Ed., *Mechanical Testing and Evaluation,* Vol 8, *ASM Handbook,* ASM International, 2000
- I.J. McColm, *Ceramic Hardness,* Plenum Press, 1990
- R.W. Rice, *Porosity of Ceramics,* Marcel Dekker, Inc., 1998
- R.W. Rice, *Mechanical Properties of Ceramics and Composites,* Marcel Dekker, Inc., 2000
- J.C. Russ and R.T. Dehoff, *Practical Stereology,* 2nd ed., Kluwer Academic, 2000

CHAPTER **10**

Qualitative Ceramography

NOT ALL CHARACTERISTICS of a ceramic microstructure are easily quantified or even need to be quantified. Properties of ceramics and many failure causes are related to the qualitative aspects of ceramography. This chapter describes microstructural characteristics that are observed rather than measured.

Morphology

Morphology (Greek: form + science) is the scientific study of the form and structure of microstructural features. The morphology of a grain or particle includes its size, shape, surface, homogeneity, transparency, color, refractive indices, and birefringence. This section is mostly concerned with descriptive grain shape.

Terms used to describe microstructures mostly apply to metals that nucleate and grow from a melt or recrystallize after heavy deformation. These terms occasionally apply to ceramics, especially secondary phases in sintered ceramics, and precipitates in glass or slag. *Acicular* grains are needle-shaped. *Laths* are grains in the shape of flat toothpicks. *Dendritic* (branching) microstructures occur when a primary phase solidifies first by extending "branches" into a eutectic melt. Most zirconia and 99.5% alumina microstructures are *equiaxed,* meaning that the length and width of the grains are about the same, and their aspect ratio is close to unity. Ceramic grains are *euhedral* or *idiomorphic* when the shape of the grain is determined by its crystal structure; that is, the shape of the grain is similar to the shape of the Bravais lattice cell. The shape of an *anhedral* grain is determined externally, by the constraints imposed by its neighboring grains. A microstructure of discrete larger grains surrounded by a matrix of smaller grains is said to be *necklaced,* such as 99.5% alumina. A microstructure containing alternating layers of coarse and fine grains is *banded. Whiskers* are fiberlike grains in which the length is approximately ten times the diameter; they are often used to reinforce other ceramics or plastics.

Phase Determination

How does one know what he is looking at when examining a microstructure with more than one phase? Many of the microstructures of oxides in Chapter 8 contain more than one oxide phase, evident from the radically different colors or gray levels between neighboring grains. The Al_2TiO_5 in Fig. 8.74(a) appears to contain at least two distinct phases. It could logically be assumed that one of the phases is β-Al_2TiO_5 and the other is Al_2O_3, TiO_2, another crystal structure (polymorph) of Al_2TiO_5, or some other Al_2O_3-TiO_2 compound. Several techniques are available for identification of phases. The determination of phase content was presented in Chapter 9. The term *interpretation of microstructure* generally begins with phase identification and ends with inferences regarding the fabrication processes that begat the microstructure, the failure mechanisms, or the expected performance of the ceramic.

The first tool in phase identification is a phase diagram, such as the Al_2O_3-TiO_2 equilibrium diagram shown in Fig. 10.1. A phase diagram shows which phase or phases are in thermodynamic equilibrium at a particular temperature and composition. The Al_2O_3 and TiO_2 mass percentages are unknown for the microstructures in Fig. 8.74(a) and (b) but some assumptions could be made simply by examining the horizontal line at room temperature in the Al_2O_3-TiO_2 phase diagram at all compositions of Al_2O_3 and TiO_2. The phase diagram in Fig. 10.1 does not show room temperature on the abscissa, but it can be assumed that the three phases stable below the 1705 °C eutectic are stable down to room temperature. The three best candidates are $\beta Al_2O_3 \cdot TiO_2$, stable at all compositions except the extreme edges; Al_2O_3, stable from the left edge (0% TiO_2, 100% Al_2O_3) to approximately 44% TiO_2; and TiO_2, stable from 44% TiO_2 to the right edge (100% TiO_2). One would infer the microstructure is mostly $\beta Al_2O_3 \cdot TiO_2$ and either Al_2O_3 or TiO_2. Metastable or nonequilibrium phases, not shown on an equilibrium phase diagram, present a fourth possibility in this system. If the other phase is Al_2O_3 and the overall composition is just under 44% TiO_2, it is conceivable that some TiO_2 is present as a metastable phase, absent any other information. The original ceramic powder mixture may have contained fluxes that further complicate the situation. Fluxes are minor quantities of other compounds, such as B_2O_3, that aid the sintering process but can affect phase stability somewhat. The *Phase Diagrams for Ceramists* series of books contains thousands of ceramic phase diagrams and an explanation of how to interpret the diagrams.

A second tool in phase identification is the x-ray microanalysis capability of most scanning electron microscopes, discussed in Chapter 7. The EDS or WDS cannot identify compounds but can accurately estimate the mass and molar percentages of the elements present in a microconstituent. The phase compositions can be inferred from the mole fractions of the elements. In Fig. 7.15, the WDS determined the mole fractions of zirconium (15 mol%), silicon (14%), and oxygen (71%) roughly equivalent to their proportions in $ZrSiO_4$.

A third tool in phase identification is x-ray diffraction (XRD), which identifies compounds rather than elements, as in ASTM D 934. X-ray diffraction is usually the best way to accurately identify the major phases in a specimen but requires a powdered sample and a laborious analysis. The TEM, described in Chapter 7, can generate an electron diffraction pattern of a single grain. The method of XRD matrix flushing described in Ref 2 quantifies the major phases in a specimen. X-ray diffraction is explained in great detail in a number of books, including a recent one by Suryanarayana and Norton.

Chemical microscopy, briefly mentioned in Chapter 7, is a fourth tool in phase analysis. In chemical microscopy, compounds in a granulated substance are identified by their refractive indices. The technique can be slow and tedious, especially for the novice microscopist, and is limited to

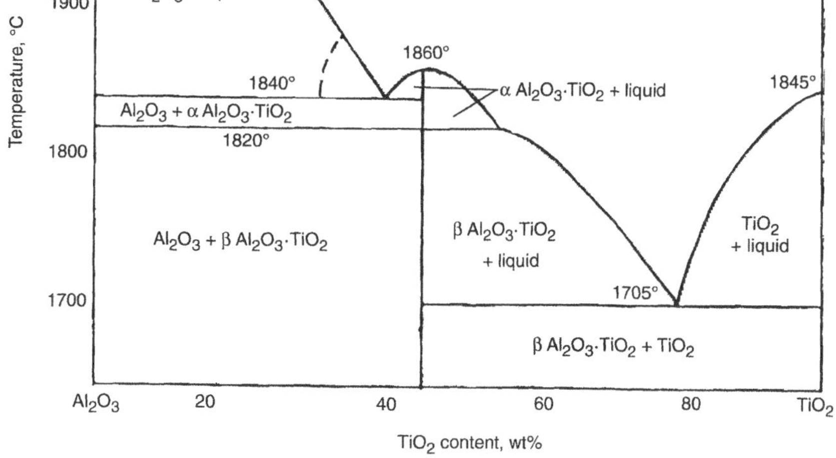

Fig. 10.1 The Al_2O_3-TiO_2 equilibrium phase diagram. Source: Ref 1

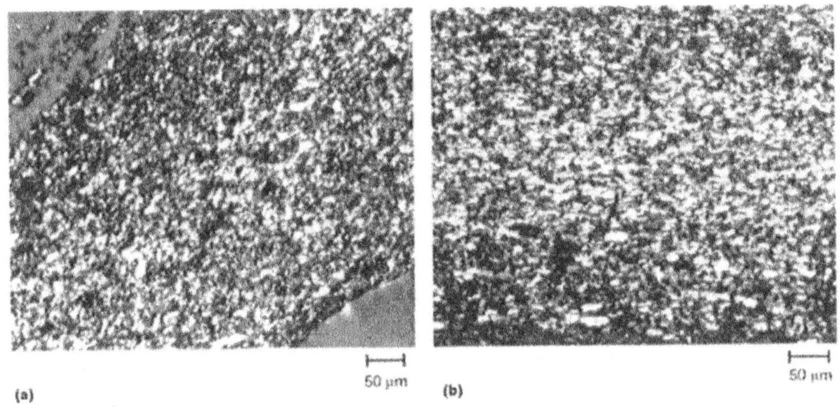

Fig. 10.2 A 99% Al_2O_3 extruded tube that has preferred orientation. (a) The tube axis is approximately 45° from the vertical and horizontal axes of the photo. (b) The same tube rotated 45°. The tube axis is parallel to the horizontal axis of the photo.

translucent compounds. Chemical microscopy can be used to identify a single particle.

Preferred Orientation

Preferred orientation is the alignment of crystallographic planes of a polycrystalline solid in some particular direction, such as the direction of maximum strain. Preferred orientation is also called texture and can be detected in ceramics either by XRD or transmitted-light microscopy. Preferred orientation is often undesirable in structural members that are subject to stresses in more than one direction, because the ceramic may be significantly weaker in one direction than another. Preferred orientation could be desirable in ferroelectric or ferromagnetic ceramics, so as to maximize the alignment of domains. Figure 10.2(a) and (b) show two views, 45° apart, of textured alumina.

The procedure for observing preferred orientation in a thin section is as follows:

1. Prepare a thin section of alumina or other noncubic, translucent ceramic as described in Chapter 6.
2. View the specimen in fully polarized transmitted light, which requires both the polarizer and analyzer to be in place and 90° apart. A full-wavelength compensator can be used in the microscope to clarify the image with color.
3. Rotate the stage to determine the angle of maximum transmitted intensity. Rotate the stage again 45° to the angle of minimum transmitted intensity or maximum extinction. The difference between the minimum

and maximum intensities is indicative of the degree of preferred orientation. A ceramic that does not have preferred orientation has the same transmitted intensity in a multigrain field at all angles of stage rotation. A textured ceramic exhibits a sharp difference between an all-dark field of view versus an all-bright field when rotated 45°. When using the full-wavelength compensator, the textured field changes from nearly all one color to nearly all a different color when the stage is rotated 45°.

4. If the microscope is equipped with a camera exposure meter, the minimum and maximum intensities, or their ratio, can be measured to obtain a quantitative measure of preferred orientation.

Fractography

Fractography is the study of fracture surfaces, usually for the purpose of determining the origin of a fracture and root cause or mechanism of a failure. The origin of a fracture may be something as obvious as a large notch, hole, corner, or dimple on an external surface of a ceramic device, or it could be something microscopic, such as a pore or oversized grain located at a point of high mechanical stress, or a sharp change in thermal gradient. The stereomicroscope and SEM are both useful for observation of fracture surfaces. An in-depth discussion of fractography and failure analysis of ceramics is beyond the scope of this book, but these topics are related to ceramography and are presented here in brief.

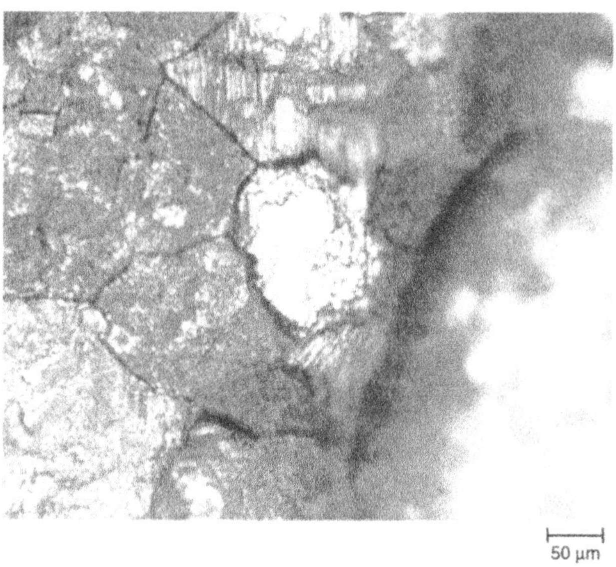

Fig. 10.3 Replica of a crack surface of as-fired $MgAl_2O_4$

Failure origins and mechanisms in fractured devices can often be determined from examination of the fracture surfaces. Features such as "gull wings" indicate the direction of the crack propagation. Cracks propagate in tension and typically originate at a large pore, an oversized grain, a foreign particle (inclusion), or a sharp notch.

Microscopic examination of a fracture surface can be done at low magnification with a stereomicroscope. Higher magnifications are obtainable from a replica in a light microscope, such as Fig. 10.3. The highest magnifications can be obtained from either the specimen or its replica in a SEM. The SEM can also generate stereo pairs (discussed in Chapter 7) of micrographs for three-dimensional viewing.

A high-strength 96% alumina displays transgranular fracture in Fig. 10.4(a). Rapid crack growth, such as in a single overload failure, is usually transgranular (Ref 3). The grain boundaries are difficult to see. A liquid phase aided the densification then crystallized into dense secondary phases. A very low-strength 96% alumina in Fig. 10.4(b) shows an intergranular fracture. Most of the grain boundaries are sharply defined. No liquid phase was present during sintering, resulting in numerous large flaws. The 96% alumina in Fig. 10.4(c) was stronger than the ceramic of Fig. 10.4(b), and the fracture surface is mixed but more intergranular than transgranular. The liquid phase was too viscous to adequately remove the

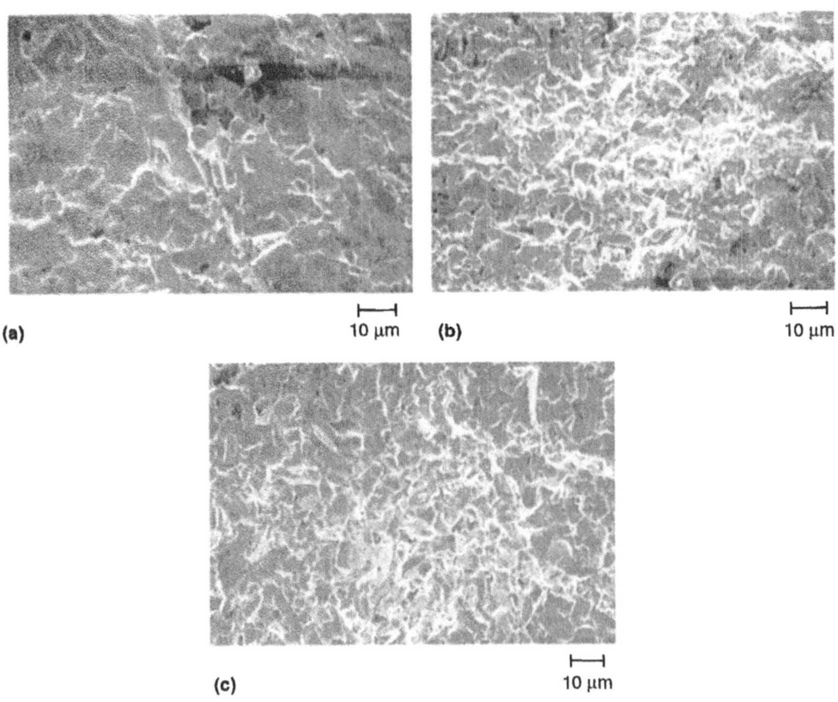

Fig. 10.4 Fracture surfaces of 96% Al2O3 MOR bars. (a) Transgranular fracture surface in a high-strength 96% Al_2O_3 MOR bar. (b) Intergranular fracture surface in a low-strength 96% Al_2O_3 MOR bar. (c) Mixed but mostly intergranular fracture surface in a low-strength 96% Al_2O_3 MOR bar

flaws, even though the density was the same as the ceramic in Fig. 10.4(a). All three ceramics failed due to a single overload, namely, the four-point bending stress from a modulus of rupture test. Rice's book summarizes research on fracture mode as a function of grain size and other microstructural characteristics for various ceramics.

The 96% aluminas in Fig. 10.5 and 10.6 failed from single overloads in MOR tests, but the fracture origin in each case was a diamond pyramid hardness indentation on the tensile surface. The indentation was more severe, that is, larger or sharper, than any of the fabrication flaws in the tensile region. The fracture path in Fig. 10.6 includes a sizable flaw in addition to the indentation. Three indentations on the tensile surface of a MOR bar provide sites where fractures are likely to originate in indentation toughness tests. A fracture path in an indentation toughness bar that does

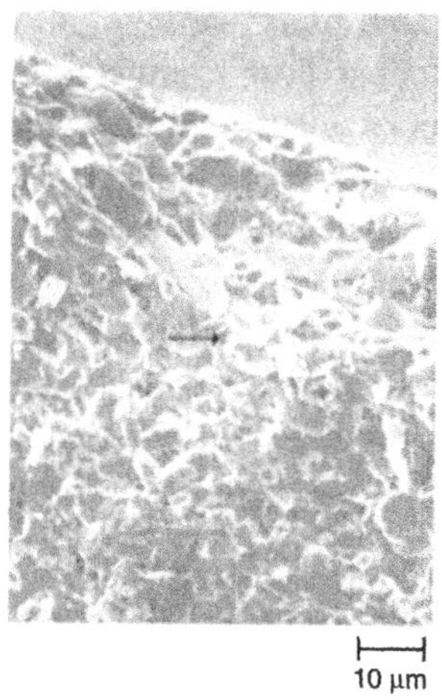

Fig. 10.5 Fracture origin at a diamond pyramid hardness (DPH) indentation in a 96% Al_2O_3 fracture toughness bar. The arrow points to the peak of the pyramid-shaped indentation. The fracture was intergranular, because the grain boundaries are clearly visible in the fracture surface.

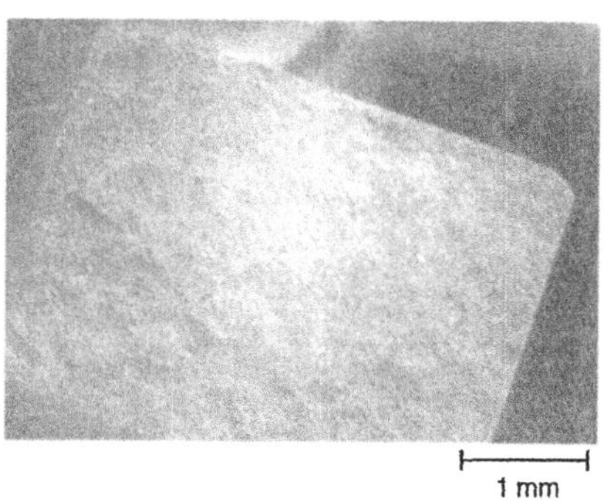

Fig. 10.6 Fracture origin at a DPH indentation in a different 96% Al_2O_3 fracture toughness bar. The indentation is in the center of the top edge. The fracture path includes the defect (dark elongated void) near the left edge.

not include an indentation is an indication that the fabrication flaws are more severe than the indentations.

The silicon carbide heating element in Fig. 10.7 failed when a furnace was slowly heated to 800 °C and slowly cooled back to room temperature. The element had been heated and cooled many times prior to the failure. The failure occurred approximately 1 cm inside the hot zone of the furnace, that is, approximately 1 cm from where the element goes into the furnace wall refractory. The failure site is also the transition zone between α-SiC and β-SiC. As the crack propagated, the continuous cross section became smaller, and the density of the electric current passing through it increased. Just before the failure, the current density was so high that it melted the SiC that had not yet broken.

Stress-corrosion cracking in a ceramic device is caused by the simultaneous presence of a corrosive environment and a tensile stress field. Stress-corrosion cracking and other forms of slow crack growth are usually intergranular (Ref 3). The slow fracture in alumina is controlled by dissociative chemisorption of electron donors and proton donors at the aluminum-oxygen bonds at the crack tips (Ref 4). Stress-corrosion cracking can also occur in fatigue loading. Figure 10.8 shows a possible stress-corrosion crack in an alumina section.

Reference texts on fractography include *Fractography,* Volume 12 in the *ASM Handbook* series. Reference 5 has a chapter on failure analysis of ceramics. Wulpi's book includes descriptions of failure mechanisms in brittle materials.

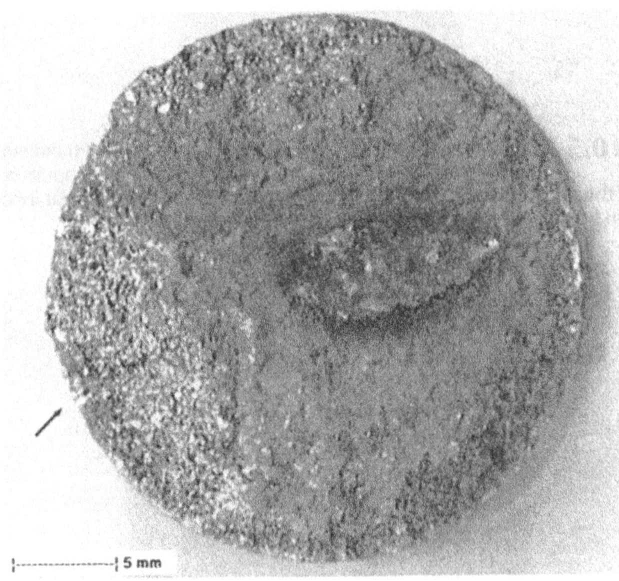

Fig. 10.7 Fracture surface in a SiC furnace element that failed from thermal fatigue and oxidation. The primary crack front originated from the 8:00 position, near the arrow. A secondary crack originated from the 11:00 position.

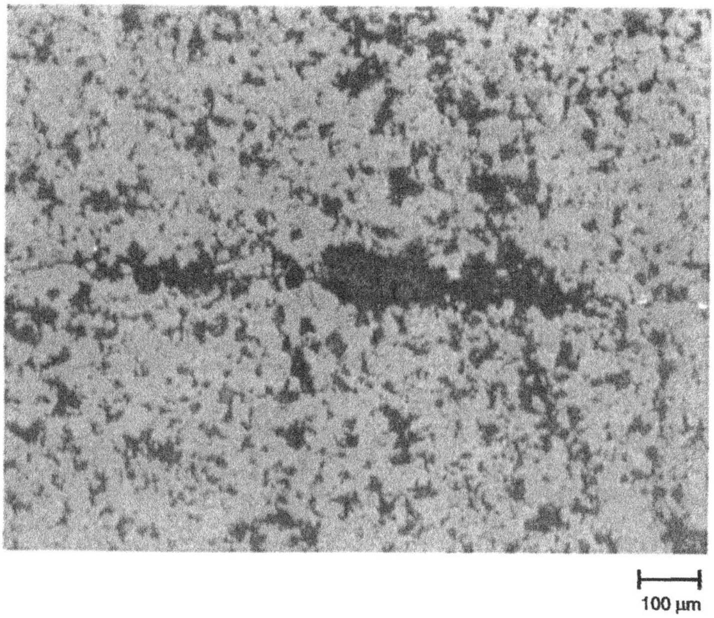

Fig. 10.8 Fracture cross section in a 99.5% Al_2O_3 flow rate sensor lining. The cracks originated at the rims of holes on the outer surface (left edge of photo), where metal inserts thermally expanded. Hoop stresses propagated the cracks inward. Corrosion by the metered liquid may have contributed to the crack growth.

Artifacts

Artifacts are artificial microstructural features that do not exist in the true microstructure but are created during the preparation and imaging steps. Artifacts are undesirable and must be avoided. Some artifacts are quite obvious, while others are easily mistaken for true microconstituents.

Pluckouts, also called *pullouts,* occur among pores, and the two can be confused for one another in many ceramics. Pores are usually rounded, smooth, and randomly spaced. Pluckouts tend to be sharp-cornered, irregularly shaped, and concentrated away from specimen edges. Pluckouts that form a line indicate a scratch that was not fully removed during grinding and polishing. *Comet tails* are streaks that frequently follow pluckouts in the direction of abrasion. Porosity can be verified by density measurements that are independent of microstructural preparation, such as the suspended-weight methods of ASTM C 20 and C 373 based on Archimedes' principle, or pycnometry. Underfired and extremely brittle ceramics are generally more susceptible to pluckouts than are fully densified and tougher ceramics.

Table 10.1 Causes of ceramographic artifacts and their remedies

Artifact	Description	Likely source	Remedies
Pluckout or pullout	Jagged-edged void	Grinding	• Use free abrasives rather than fixed. • More thorough polishing
		Thermal etching	More gradual heating and cooling
		Underfired ceramic	• Higher sintering temperature • Longer sintering time • Addition of a sintering aid or flux, e.g., B_2O_3
Crack	Split at sharp corners or notches	Compression mounting	• Use castable resin. • Trim away sharp corners and notches before compression mounting.
		Thermal etching	More gradual heating and cooling
Comet tail	Dark streaks out of pluckouts	Polishing	More thorough polishing and cleaning. Polish in more than one direction.
Scratch	Long groove or series of pluckouts	Grinding	• Increase grinding time. • Increase grinding force.
		Polishing	• Frequently replenish free abrasive and lubricant. • Replace polishing paper to remove contaminants.
		Chemical etching	Thermally etch if possible—chemical etchants attack scratches.
Foreign particle	Particle not in the polished plane	Residue from encapsulation burnout	Ultrasonicate after burnout.
		Thermal etch furnace	Use a covered crucible.
		Atmospheric dust	Ultrasonicate right before viewing. Store in a clean cabinet.
		Residual abrasive particle	Swab and ultrasonicate after each grinding and polishing step. See Subroutine 4.1.
Deep corroded region	Large pits in etched surface	Overetching	Use a less severe etch, e.g., less time, lower temperature, lower concentration of reagent, less electrolytic current, etc.
Shallow corroded region	Discolored or clouded etched surface	Chemical etching	Lightly polish with 0.05 μm γ-Al_2O_3 or colloidal silica for a few seconds. Clean the specimen as in Subroutine 3.1.
Edge rounding	Encapsulation-specimen interface out of focus	Grinding and polishing	Less intensive grinding and polishing
		Encapsulation	Use a resin with abrasion resistance closer to that of the specimen.
Water spots	Large, round, translucent overlay or stain	After cleaning in liquid	Use softer (less mineral content) water. Deionized or distilled water is preferred to tap water. Dry the specimen under a heat gun.

Most artifacts can be avoided by improved practice on the part of the ceramographer, such as choice of encapsulation, thorough polishing absent of contamination, and thorough cleaning of specimens after each stage of preparation. Table 10.1 gives some examples of artifacts, their causes, and remedies.

REFERENCES

1. E.M. Levin, C.R. Robbins and H.F. McMurdie, Ed., *Phase Diagrams for Ceramists,* 2nd ed., The American Ceramic Society, 1969, p 123
2. F.H. Chung, Quantitative Interpretation of X-Ray Diffraction Patterns of Mixtures, Part I: Matrix-Flushing Method for Quantitative Multicomponent Analysis, *J. Appl. Crystallogr.,* Vol 7, 1974, p 519–525

3. R.W. Rice, *Mechanical Properties of Ceramics and Composites,* Marcel Dekker, Inc., 2000, p 665–670
4. T.A. Michalske, B.C. Bunker, and S.W. Freiman, Stress Corrosion of Ionic and Mixed Ionic/Covalent Solids, *J. Am. Ceram. Soc.,* Vol 69 (No. 10), 1986, p 721–724
5. D.W. Richerson, *Modern Ceramic Engineering: Properties, Processing, and Use in Design,* 2nd ed., Marcel Dekker, Inc., 1992, p 680–730

SUGGESTED READING

- *Fractography,* Vol 12, *ASM Handbook,* ASM International, 1987.
- *Phase Diagrams for Ceramists* (five volume series), The American Ceramic Society, 1964, 1969, 1975, 1981, 1983
- R.W. Rice, *Mechanical Properties of Ceramics and Composites,* Marcel Dekker, Inc., 2000
- C. Suryanarayana and M.G. Norton, *X-Ray Diffraction, A Practical Approach,* Plenum Press, 1998
- D.J. Wulpi, *Understanding How Components Fail,* 2nd ed., ASM International, 1999

CHAPTER 11

Image Analysis

IMAGE ANALYSIS is a means of automated evaluation of microstructures and other stereological images. An image analyzer consists of a computer capable of digitizing an image and a camera connected to a microscope. Ideally, an image analyzer increases the speed of microstructural measurements and eliminates human bias through automation and unified methods. The essence of an image analyzer is its software and customized computer board. Virtually all the other components—computer, microscope, camera, monitor, and so on—are off-the-shelf items. Many of the larger metallographic equipment and microscope manufacturers now offer an image analysis system.

An image analyzer operates on a digital image from a digital camera or scanner. An analog image from a charge-coupled device camera can be digitized by the frame-grabber board inside an image analyzer computer. A black-and-white digital image consists of an array of *pixels* (picture elements) that have *x-y* coordinates and a gray level. The human eye can distinguish only approximately twenty gray levels, but the digital image is divided into as many as 256 (2^8) gray levels, from zero (all black) to 255 (all white). A black-and-white digital image may be *pseudocolored* by an image analyzer; that is, colors with digital dimensions are substituted for specific gray levels.

A color digital image has the dimensions of hue, luminance (or intensity, value, or lightness), and saturation (HLS) instead of gray level and requires a color camera. An alternative to HLS space is red, green, blue (RGB). Digital colors, which are also used in Java programming for HyperText Markup Language web-page graphics, have three dimensions in RGB space. A color pixel has 24 bits, or 2^{24} ($2^8 \times 2^8 \times 2^8$, or approximately 16 million) color combinations. An all-black pixel would have the color coordinates (0, 0, 0) in addition to its *x-y* coordinates in RGB space, an all-white pixel (255, 255, 255), an all-red (255, 0, 0), and so on.

Most image analyzers can be used manually. That is, the user traces the grains in a displayed image with a mouse or electronic stylus. An automatic image analyzer uses a macro—an algorithm refined into a sequence of the image analyzer menu functions and commands from a programming language—to process and measure an image. Some image analyzers use a hybrid of the *C* programming language, which should not be confused with the more common *Visual C++* (Microsoft Corp.). The algorithm includes five basic steps:

1. Prepare the image analyzer for measurement.
2. Capture an image from a microscope or other source.
3. Prepare the image for analysis.
4. Measure objects of interest in the image.
5. Plot the measured data and print a report.

Algorithm

Menu commands are unique to each brand of image analyzer, but the functions they represent are largely the same from one brand to the next. The vernacular of image analysis also varies among manufacturers and authors. The grayscale algorithm is described subsequently in greater detail, with emphasis on grain size measurement. Some steps are combined into one operation by some image analyzers or do not apply for other reasons. Steps 3(b–d) and (f) describe electronic means of optical etching.

Step 1: Prepare the image analyzer for measurement:

a. Clear all of the image memories or bit planes.
b. Clear all overlays. An *overlay* is a transparent binary image superimposed on a grayscale image, much like an acetate overlay on paper or the layers in an engineering drafting program. Overlays can contain text, a grayscale histogram, scale bars, or reconstructed grain boundaries, for example. The overlay can be cleared or edited independent of the underlying image.
c. Calibrate the camera magnification, as necessary, with a stage micrometer if the camera is mounted on a microscope. Use a ruler or other scale for calibration if the camera is mounted on a macroviewing stand or if the image comes from a scanner. Magnification calibration is discussed in Chapter 7. The image analyzer may require two separate calibrations, one in the x-dimension and the other in the y-dimension. The aspect ratio of the image is not necessarily 1 to 1.
d. Configure the system and declare and initialize the variables and vectors, as necessary. Some variables may need to be reset to zero or some other initial value each time the program is run.
e. Select a look-up table (LUT). A LUT is a means of mapping or transforming the gray levels of an image to another set of gray levels or

pseudocolors. Binarization, discussed in step 3(e), is an example of a LUT applied to an image. Image rotation and binary inversion are also types of LUTs. Most image analyzers default to a neutral or 1 to 1 LUT that does not transform the captured image at all.

Step 2: Capture an image from a microscope or other source:

a. Turn the camera on and adjust its focus so that it is in focus when the oculars of the microscope are in focus.
b. Optimize the image input by adjusting the microscope, the illumination intensity, and the gain and offset controls of the camera. Use Köhler illumination, described in Chapter 7.
c. Read the image into the first image memory of the computer. The image analyzer computer has a frame-grabber board that converts the analog video signal into a digital array of pixels. A digital camera or scanner can also be used with most image analyzers. An etched spinel microstructure is shown in Fig. 11.1(a) as an example.

Step 3: Prepare the grayscale image. Clean the binary image:

a. Normalize the gray levels to improve contrast. That is, spread the gray levels out over the entire range of gray levels, from zero (all black) to 255 (all white). Save the normalized image in the second image memory, leaving the original image unchanged.
b. Correct the shading. The intensity of the image is ordinarily not uniformly distributed, even with Köhler illumination. One way the image analyzer corrects the shading is by subtracting an electronically defocused, highly filtered image from the focused image. The electronically defocused image has the same nonuniform intensity distribution as the focused image, such that excessively bright spots and dark spots are removed. Restore the corrected image to normal overall brightness in the third image memory. The corrected, normalized image is usually the best one for publication in the final report. Add scale bars as an overlay before printing this image.
c. Reduce the noise and remove the undesirable features. One way to reduce the noise in an image is to apply a filter. A 3×3 median filter calculates the median gray level of every nine-pixel square array or kernel, then replaces the center pixel with the median value. Noise in the form of extreme gray values is deleted, but some fine details of the image may also be lost. Save each change in a subsequent image memory. The spinel from Fig. 11.1(a) is shown in Fig. 11.1(b) after a shading correction and 3×3 median filter.
d. Enhance the contrast between contiguous desirable features. An edge-sharpening operation increases the contrast between grains and boundaries by creating a steep gray-level gradient where grains and boundaries intersect. Sputter coating the polished surface with

a thin layer of gold can enhance the contrast between grains and boundaries, as described in Chapter 7. The spinel in Fig. 11.1(c) is the same as Fig. 11.1(b) after application of an edge-sharpening filter.

e. *Binarize, segment,* or *discriminate* the image into objects and background by choosing the lower and upper limits of a gray-level threshold. Pixels with gray levels between the limits become objects, and pixels outside the limits become background. The objects are assigned a binary gray level of one and appear white. The background is assigned a binary gray level of zero and appears black. For grain size measurement, set the threshold values so that the grains become objects and the boundaries, pores, glass, and everything else become background, as in Fig. 11.1(d). The binary image usually contains some noise in the form of one- or two-pixel "objects" that can be deleted in the next step or ignored in the measurement. A common problem in the automated image analysis of etched ceramic microstructures is that the gray levels of grains and grain boundaries often overlap each other. As a result, some of either the grains or the grain boundaries are lost in the binarization step.
f. Clean the binary image by deleting the noise particles. One way to delete pores within grains is to invert the objects and background, as in Fig. 11.1(e), so that the pores become discrete objects. Identify and delete undersized objects, say, less than 5 pixels. Invert the binary image again so that the grains are the objects once more. Erosion and dilation operations are one way to close disconnected grain boundaries, as in Fig. 11.1(f). Pores and cracks may be adversely affected by dilation and erosion.
g. Reconstruct the grain boundaries, if the grain size is being measured. The more sophisticated image analyzers have proprietary software or firmware that reconstructs grain boundaries automatically, with some degree of error. The image in Fig. 11.1(g) is the image in Fig. 11.1(c) after an operation that compass-filtered, binarized, and reconnected the boundaries.
h. Manually edit the image, if necessary. Draw in incomplete grain boundaries and delete undesired objects or holes, as in Fig. 11.1(h).

Step 4: Measure objects of interest in the image:

a. Choose the measurement parameters, for example, mean diameter, area, perimeter, minimum diameter, maximum diameter, and shape factors. These parameters may be grouped into a vector.
b. Select the objects or features to be measured. Objects that intersect the image boundary should be excluded. The image boundary is also called the region of interest.
c. Set the conditions for measurement, such as the exclusion of objects outside a size range. For example, objects of four pixels or fewer or

objects less than 0.5 μm in diameter should be excluded if they are beneath the resolution limit of the optical instrument. The resolution of an image analyzer is always limited to one pixel.

d. Define the data file or spreadsheet for measurement data.
e. Begin the measurement function.
f. Return to step 1(a) for another field when all the objects have been measured or continue with step 5. The cumulative data from several nonoverlapping fields on a specimen yield more reliable measurements than the data from just one field.

Step 5: Plot the measured data and print a report after all the fields on a specimen have been measured. Some image analyzers draw plots, tables, and histograms with the image analysis software, while others export all measurements to an Excel (Microsoft Corp.) or Quattro (Corel Corp.) spreadsheet.

a. Plot a histogram of the diameter distribution, if grain size or pore size is being measured. A grain size distribution histogram is shown in Fig. 11.1(i) and (j).
b. Print a statistical table of all measured data and note any measurement conditions set in step 4(c). The mean, maximum, minimum, and standard deviation of each measurement parameter are usually desirable.
c. Plot a scatter diagram to show a correlation between variables, such as shape factor as a function of diameter.
d. Print a report of relative area measurements, if porosity content or second-phase content was measured.
e. Print the normalized, corrected image with scale bars on the same sheet as the histogram and table, if possible.
f. Return to step 1(a) for another specimen or exit to the computer operating system.

For porosity or secondary-phase content, the algorithm is largely the same as the grain size algorithm. The main differences are that steps 3(f–h) are unnecessary. In step 3(e) segment the pores or particles of the secondary phase as objects and the grains of the primary phase as the background. A porosity measurement image is shown in Fig. 11.2(a) and the pore size distribution histogram in Fig. 11.2(b).

Critical Aspects

Three critical aspects of automatic image analysis are image input, calibration, and binarization.

An image analyzer can only do so much to improve an image, and it cannot detect features that are not visible in the image. Quality of micro-

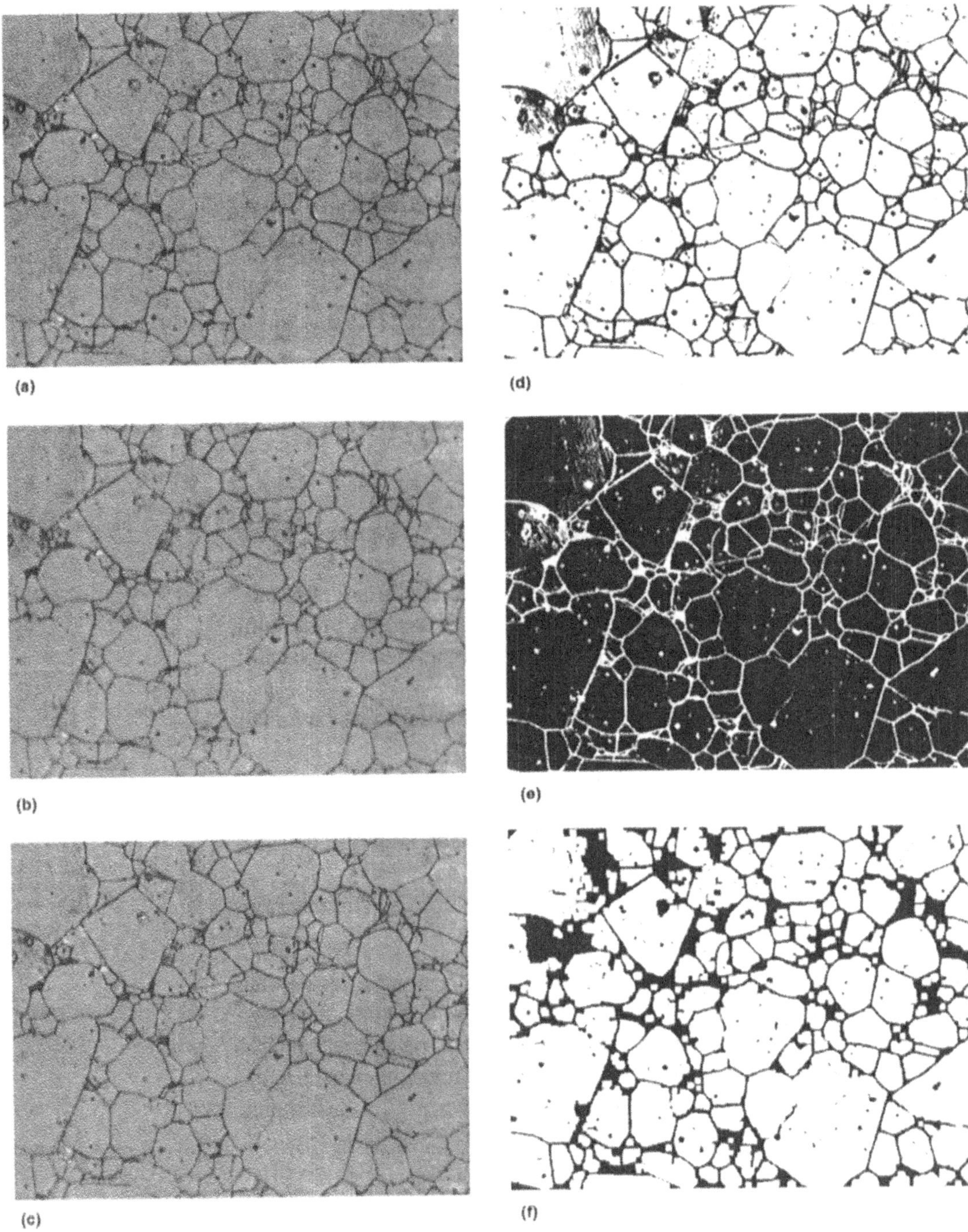

Fig. 11.1 Image analysis of the spinel in Fig. 8.90. (a) Grayscale bitmap. (b) The grayscale image from (a) after background illumination correction and a 3 × 3 median filter. (c) The grayscale image from (b) after use of an edge-sharpening filter. (d) The image in (c) after binarization. The grains are objects, and the boundaries are background. (e) Binary inversion of the image in (d). The grain boundaries and pores are now the objects. (f) The binary image from (d) after three erosion and dilation steps to reconnect grain boundaries. Some boundaries remain disconnected. (g) The binary image from (c) after use of a compass filter, threshold, and watershed to reconnect grain boundaries. (h) The binary image from (e) after manual editing and a second inversion. (i) Histogram of the grain size distribution of the spinel in (g). The distribution is exponential rather than normal (Gaussian). (j) Histogram of the same grain size distribution as in (i) but with a logarithmic *x*-axis

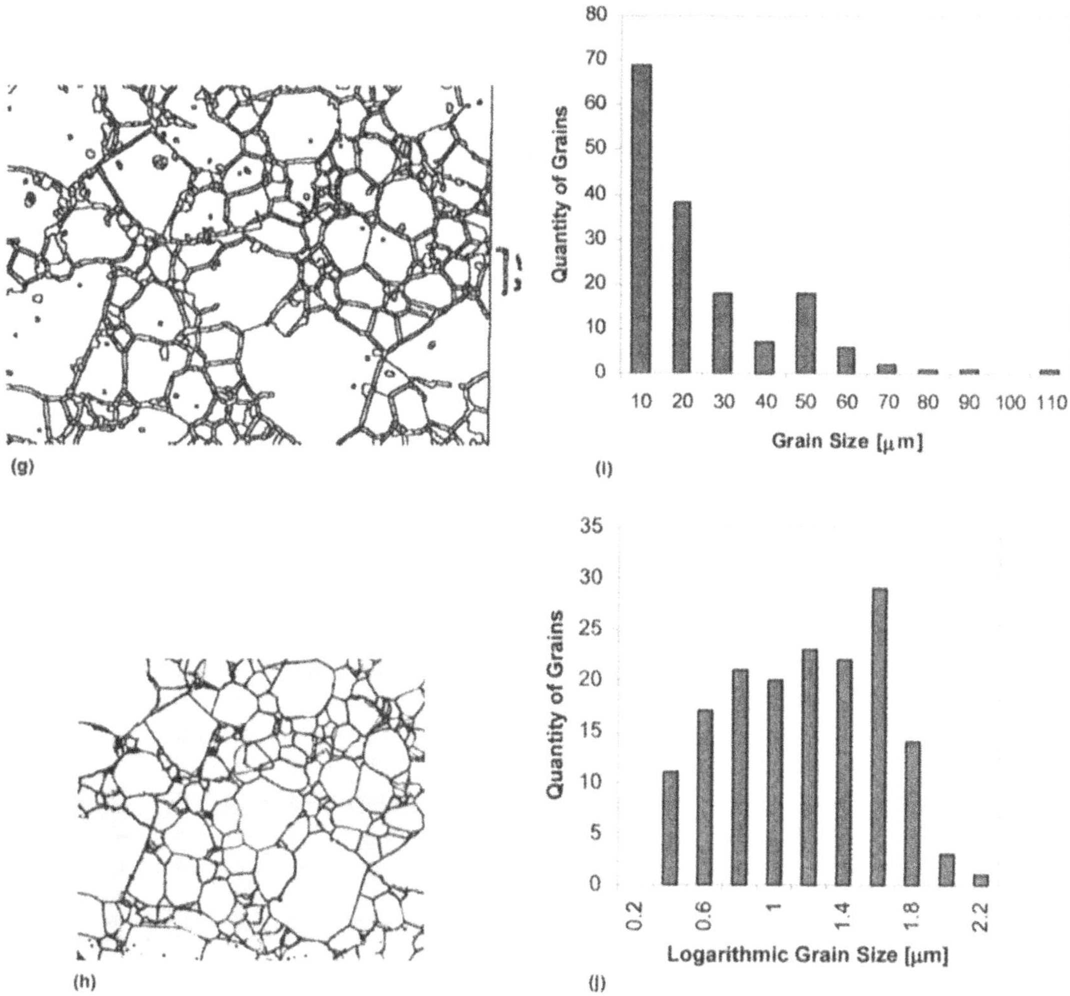

Fig. 11.1 (continued) (g) The binary image from (c) after use of the compass filter, threshold, and watershed to reconnect grain boundaries. (h) The binary image from (e) after manual editing and a second inversion. (i) Histogram of the grain size distribution of the spinel in (g). The distribution is exponential rather than normal (Gaussian). (j) Histogram of the same grain size distribution as in (i) but with a logarithmic *x*-axis.

structure, due to preparation techniques (Chapters 3 to 5) and microscope settings (Chapter 7), strongly affect the image analysis results (Ref 1). The computer programmers' truism, "Garbage in, garbage out!" certainly applies to image analysis.

Image analysis results are misleading and useless if the image analyzer is not calibrated or calibrated incorrectly. The measurement of area fractions, such as porosity or second-phase content, does not require calibration, but measurement in real units, such as microns or millimeters, does require calibration. Use reliable standards such as a stage micrometer to calibrate the image analyzer. Recalibrate the image analyzer anytime the video camera is moved or adjusted, even though the calibration change may be minor. Keep in mind that the magnification seen by the image analyzer is not necessarily the same in the *x*- and *y*-directions, and that the magnification seen on the video monitor is not necessarily the same as the magnification in the microscope oculars or photograph port.

Binarization, also called discrimination, segmentation, or thresholding, divides a grayscale image into two discrete gray levels, usually black (gray level = 0) and white (gray level = 255 but set to 1 by a LUT and pseudocolored white). Grain separation is highly dependent on the quality of the binarized image. Gray levels of grains, boundaries, and second-phase particles often overlap in ceramic microstructures. Thus, binarization might not separate all contiguous grains. Image analyzers are equipped with a mouse or electronic stylus that allows one to manually edit a binary image, that is, draw in incomplete or missing grain boundaries. The disadvantage of manual editing is that it introduces operator bias, something that automated image analysis is intended to prevent. The shading correction, median filter, and edge-sharpening steps on the grayscale image help optimize

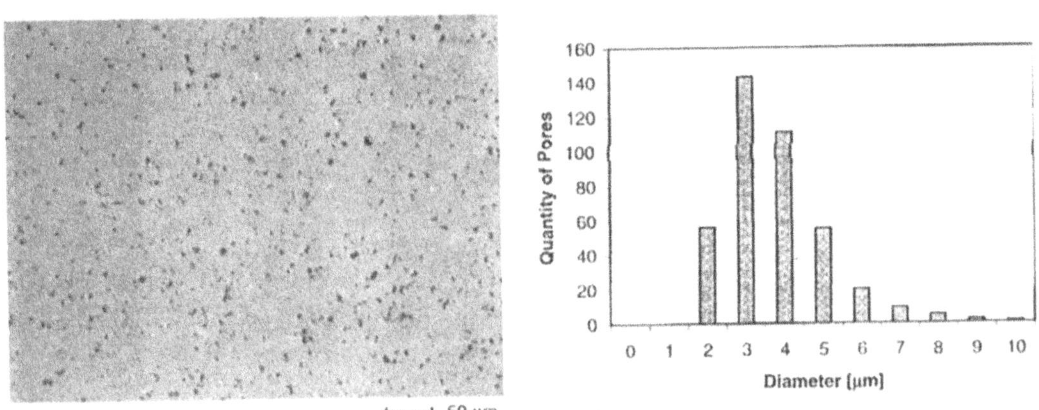

(a) (b)

Fig. 11.2 Image analysis to determine porosity. (a) Unetched 96% Al_2O_3 after background illumination correction, a 3 × 3 median filter, and an edge-sharpening filter. The porosity was measured as 4.7%. (b) Histogram of the pore size distribution of the pores in (a). The average pore diameter was measured as 4.4 μm. Pores smaller than approximately 2 μm were too small (i.e., ≤1 pixel) for measurement at this magnification.

the binarization step. The noise-removal step after binarization prepares the binary image for accurate measurement.

Color segmentation is more complex than grayscale binarization. The image is divided into two red, two green, and two blue levels.

Measurements

Image analyzers can make a number of measurements on one image. The measurement menu consists mostly of basic geometric quantities, for example, lengths (diameter, perimeter, elliptical axes, etc.), areas, and angles, while some systems allow for user-defined quantities.

The microstructural quantities of interest, such as porosity or grain size, are enhanced in the refinement steps and selected in the measurement steps. Binarization thresholds are chosen that emphasize the difference in gray levels between pores and matrix or between grains and boundaries. Image manipulations, such as erosion and dilation, can thin grain boundaries until adjacent grains are separated by as little as one pixel width. Porosity is measured as area fraction, as in Eq 1. Grain size may be defined in several ways, including the average length of a set of 24 chords drawn through the centroid of a grain in 15° increments from 0 to 360°, or the diameter of a circle with the same area as the grain, as in Eq 2:

$$P = \frac{100\,A_P}{A_{total}} = \frac{100V_p}{V_{total}} \qquad \text{(Eq 1)}$$

$$G = d_{eqv} = \sqrt{\frac{4A_G}{\pi}} \qquad \text{(Eq 2)}$$

where P is porosity (%); A_P is the area of the pores (pixel); A_{total} is the total area of the microstructure in the field seen by the image analyzer (pixel); V_P is the volume of the pores; V_{total} is the total volume of the ceramic; G is the grain size (μm); d_{eqv} is the "equivalent" diameter of a circle of the same area as the grain (μm); and A_G is the area of the grain being measured (μm^2).

Noise in the binary image occurs as white spots that are often 1 or 2 pixels in any dimension. This noise can be eliminated either by image-processing functions that delete small objects from the binarized image or measurement conditions that preclude objects beneath a specified size. Noise that is included in the measurement of grain size tends to skew the histogram such that the mean and median are less than their actual values. In general, measurement conditions should be set to preclude any object size that is beneath the resolution of the microscope to which the image analyzer is connected. A measurement condition that is set too high might exclude some legitimate, very small grains or pores along with the noise. The uncertainty of an image analyzer measurement is at least 1 pixel; therefore a 2-pixel object has an error of ±50%.

The grain size histogram usually shows an exponential distribution rather than normal or Gaussian, as in Fig. 11.1(i). That is, the classes containing the smallest grains have much larger quantities of grains than the classes containing the larger grains. When the *x*-axis or abscissa is converted to a logarithmic scale, as in Fig. 11.1(j), the classes containing the smaller grains are spread out while the classes containing the larger grains are consolidated.

ASTM procedures pertaining to image analysis, such as E 1245 and E 1382, are listed in Appendix A. Some image analysis reference texts are listed at the end of this chapter.

Digital Images

Image analyzers generate and operate on digital images. A number of digital image formats are in use, often unique to a particular application or software product. Some of the more common image formats are described here in brief.

The most basic image files are *bitmaps,* which end with the file name extension .BMP. Bitmaps contain an *x-y* array of pixels, with a gray level or digital color at each pixel. Because they are not compressed, bitmaps tend to be very large image files. A typical bitmap is 512 pixels wide and 480 pixels high, a total of 245,760 pixels. A high-resolution bitmap might have 1024 × 960 pixels and require a full megabyte of memory. Uncompressed bitmaps are generally more suitable than compressed images for image analysis.

Files that end with .JPG or .JPEG are images in the Joint Photographic Experts Group format. The JPEG ("jay-peg") files are bitmaps that have been compressed to minimize the amount of disc space they occupy and are widely used in web pages. The JPEG compression algorithm uses discrete cosine transformation, which results in a loss of least-significant data from the original image. The loss of data is not necessarily a disadvantage, and the JPEG format is especially suitable for color photographs. The 24-bits/pixel (approximately 16 million colors) format, created by C-Cube Microsystems, Inc., is in the public domain.

Image files that end with .TIF or .TIFF are bitmaps in Tagged Image File Format. The TIFF format was developed jointly by Aldus Corp. and Microsoft Corp. but is now owned by Adobe Systems, Inc.

Files that end with .GIF are compressed bitmaps in Graphics Interchange Format, a copyright of CompuServe, Inc. The Lempel-Ziv-Welch GIF compression algorithm, patented by Unisys Corp., retains all information from the original image but is limited to 256 colors (8 bits/pixel).

The GIF format is well suited to line drawings and similar images with sharp boundaries. Portable Network Graphics (.PNG) files are similar to GIF files, but the format is not patent-protected. The PNG ("ping") files, similar to JPEG, have 24-bit color.

Image files that end with .WMF are Windows Metafiles. Metafiles are often generated by the export command of programs that generate plots or engineering drawings. Metafiles can be opened, edited by, and saved in Windows-based applications, such as word-processing documents and spreadsheets.

REFERENCE

1. R.E. Chinn, Grain Sizes of Ceramics by Automatic Image Analysis, *J. Am. Ceram. Soc.,* Vol 77 (No. 2), 1994, p 589–592

SUGGESTED READING

- H.E. Exner and H.P. Hougardy, Ed., *Quantitative Image Analysis of Microstructures,* DGM Informationsgesellschaft GmbH, 1988
- J.J. Friel, J.C. Grande, D.Hetzner, K. Kurzydłowski, D. Laferty, M.T. Shehata, V. Smolej, G.F. Vander Voort, and L. Wojnar, *Practical Guide to Image Analysis,* ASM International, 2000
- P.J. Kenny et al., Ed., *Image Analysis and Metallography,* International Metallographic Society and ASM International, 1989
- J.C. Russ and R.T. Dehoff, *Practical Stereology,* 2nd ed., Kluwer Academic, 2000

APPENDIX **A**

ASTM Procedures Applicable to Ceramography

ASTM Procedures

ASTM No.	Title
Microstructure preparation	
B 657	Method for Metallographic Determination of Microstructure in Cemented Carbides
B 665	Practice for Metallographic Sample Preparation of Cemented Carbides
C 856	Practice for Petrographic Examination of Hardened Concrete
C 994	Practice for Ceramographic Preparation of PuO_2-UO_2 Mixed Oxide Fuel Pellets and Uranium Pellets
E 3	Methods of Preparation of Metallographic Specimens
E 407	Methods for Microetching Metals and Alloys
E 883	Guide for Metallographic Photography
E 1351	Practice for Production and Evaluation of Field Metallographic Replicas
Evaluation of microstructure	
B 276	Test Method for Apparent Porosity in Cemented Carbides
B 390	Practice for Evaluating Apparent Grain Size and Distribution of Cemented Tungsten Carbides
B 487	Method for Measure of Metal and Oxide Coating Thicknesses by Microscopical Examination of a Cross Section
B 748	Method for Measurement of Thickness of Metallic Coatings by Measurement of Cross Section with a Scanning Electron Microscope
C 664	Test Methods for Thickness of Diffusion Coating
C 968	Method of Analysis of Sintered Gd_2O_3-UO_2 Pellets
E 20	Practice for Particle-Size Analysis of Particulate Substances in the Range of 0.2 to 75 Micron by Optical Microscopy
E 112	Methods for Determining Average Grain Size
E 562	Practice for Determining Volume Fraction by Systematic Manual Point Count
E 930	Methods of Estimating the Largest Grain Observed in a Metallographic Section (ALA Grain Size)
E 1181	Test Methods for Characterizing Duplex Grain Sizes
E 1245	Practice for Determining the Inclusion Content of Steel and Other Metals by Automatic Image Analysis
E 1268	Practice for Assessing the Degree of Banding or Orientation of Microstructures
E 1382	Standard Test Methods for Determining the Average Grain Size Using Semiautomatic and Automatic Image Analysis

(continued)

ASTM procedures (continued)

ASTM No.	Title
Definitions, specifications, and practices	
E 6	Definitions of Terms Relating to Methods of Mechanical Testing
E 7	Definitions of Terms Relating to Metallography
E 175	Definitions of Terms Relating to Microscopy
E 211	Specification for Cover Glasses and Glass Slides for Use in Microscopy
E 766	Practice for Calibrating the Magnification of SEM Using NBS SRM-484
E 807	Practice for Metallographic Laboratory Evaluation
E 986	Practice for Scanning Electron Microscope Performance Characterization
E 1951	Standard Guide for Calibrating Reticles and Light Microscope Magnifications
Microhardness and superficial hardness	
C 849	Text Method for Knoop Indentation Hardness of Ceramic Whitewares
C 1326	Standard Test Method for Knoop Indentation Hardness of Advanced Ceramics
E 6	Definitions of Terms Relating to Methods of Mechanical Testing
E 18	Standard Test Methods for Rockwell Hardness and Rockwell Superficial Hardness of Metallic Materials
E 92	Test Method for Vickers Hardness of Metallic Materials
E 384	Test Method for Microhardness of Materials

Headings in the ASTM Subject Index

ASTM adds and revises standards annually. The most recent standards pertaining to ceramography are likely to be found under the following subject index headings:

- Ceramography
- Electron microscopy
- Grain size
- Hardness
- Image analysis
- Metallography
- Microscopic examination
- Microsection scanning
- Microstructure
- Morphology
- Petrography
- Micrography
- Replicas

These procedures are available in bound volumes or individually from:

ASTM International (formerly the American Society for Testing and Materials)
100 Barr Harbor Drive
P.O. Box C700
West Conshohocken, PA 19428-2959 USA
Telephone +1-610-832-9585
http://www.astm.org

APPENDIX **B**

Ceramographic Equipment Manufacturers

The companies listed subsequently are either manufacturers or private-label distributors or provide training and literature. Many of the manufacturers have regional and foreign representatives that are not listed here. The foreign offices are usually listed in the web site. More manufacturers and distributors may be listed in http://www.metallography.com.

Company	Product lines
Ceramics Correspondence Institute The American Ceramic Society (**ACerS**) 735 Ceramic Place P.O. Box 6136 Westerville, OH 43086-6136 USA Telephone +1-614-890-4700 http://www.acers.org/education/cci	Ceramics seminars and literature *Ceramic Bulletin* magazine *Journal of the American Ceramic Society*
New York State College of Ceramics **Alfred** University Continuing Education and Technology Transfer McMahon Bldg. Alfred, NY 14802-1296 Telephone +1-607-871-2425 http://nyscc.alfred.edu	Ceramics seminars Ceramic engineering education and research
Allied High Tech Products, Inc. 2376 East Pacifica Place Rancho Dominguez, CA 90220 USA Telephone 800-675-1118 http://www.alliedhightech.com	Metallographic equipment and consumables
ASTM International (formerly the American Society for Testing and Materials) 100 Barr Harbor Drive P.O. Box C700 West Conshohocken, PA 19428-2959 USA Telephone +1-610-832-9585 http://www.astm.org	Testing procedures and literature

(continued)

(continued)

Company	Product lines
Buehler Ltd. 41 Waukegan Road P.O. Box One Lake Bluff, IL 60044 USA Telephone +1-708-295-6500 or 800 BUEHLER http://www.buehler.com	Metallographic equipment and consumables Image analyzers Hardness testers Institute for Microstructural Analysis
Clark Instrument Division Sun-Tec Corp. 24371 Catherine Industrial Drive #227 Novi, MI 48375 USA Telephone +1-248-449-1188 http://www.clarkinstrument.com/	Hardness testers
Clemex Technologies Inc. 800 Guimond Longueuil, Québec, J4G 1T5 Canada Telephone +1-450-651-6573 http://www.clemex.com	Image analyzers Canadian distributor for several international metallographic manufacturers
Deutsches Institut für Normung e.V. (**DIN**) Burggrafenstrasse 6 10787 Berlin Germany Telephone +49-30-2601-0 http://www.en.din.de/	Testing procedures and literature
Excel Technologies, Inc. 99 Phoenix Avenue Enfield, CT 06082 USA Telephone +1-203-741-3435 http://www.extec.com	Metallographic equipment and consumables
FEI Company (formerly Philips Electronic Instrument Co.) 7451 Evergreen Parkway Hillsboro, OR 97124 USA Telephone +1-503-640-7500 http://www.feicompany.com	Scanning electron microscopes See also: FEI Company Building AAE P.O. Box 218 5600 MD Eindhoven The Netherlands Tel: +31 40-2766768
Federation of European Producers of Abrasives (**FEPA**) 20, Avenue Reille 75014 Paris France Telephone +33-1-45-81-25-90 http://www.fepa-abrasives.org/	Safety documents and standards for abrasives
Ernest F. **Fullam,** Inc. 900 Albany Shaker Road Latham, NY 12110 USA Telephone +1-518-785-5533 http://www.fullam.com	SEM and microscope accessories Sputter coaters
Hillquist, Inc. 35502 S.E. Fall City Snoqualmie Road Fall City, WA 98024 USA Telephone +1-425-222-6968 http://www.hillquist.com	Petrographic equipment and consumables
Hitachi, Ltd. Nissei Sangyo America, Inc. 2850 Golf Road, Suite #200 Rolling Meadows, IL 60008 USA Telephone +1-847-981-8989 http://www.nissei.com	Scanning electron microscopes See also: Instruments Hitachi, Ltd. Tokyo, Japan Telephone +813-5223-4567 http://www.hitachi.co.jp/

(continued)

(continued)

Company	Product lines
Innovative Met Products Technical cc (**IMP**TECH) P.O. Box 1110 Boksburg, 1460 Gauteng Republic of South Africa Telephone +27-11-914-4500 http://www.imp.co.za	Metallographic equipment and consumables
International Organization for Standardization (**ISO**) 1, rue de Varembé, Case postale 56 CH-1211 Geneva 20, Switzerland Telephone +41-22-749-01-11 http://www.iso.ch	Testing procedures and literature See also: Organisation internationale de normalisation
JEOL U.S.A. Inc. 11 Dearborn Road Peabody, MA 01960 USA Telephone +1-508-535-5900 http://www.jeol.com	Scanning electron microscopes See also: JEOL Ltd. Akishima, Japan http://www.jeol.co.jp
Kay Industrial Diamond Corporation 1551 NW Gateway Drive Fort Lauderdale, FL 33442 USA Telephone +1-305-973-3040	Diamond abrasives
Ladd Research Industries, Inc. 131 Dorset Lane Williston, VT 05495 USA Telephone 800-451-3406 http://www.laddresearch.com	SEM and microscope accessories
Lam Plan S.A. 7, rue des Jardins BP 15 74240 Gaillard France Telephone +33-04-50-43-96-30 http://www.lamplan.fr	Metallographic equipment and consumables
Lapmaster International Ltd. 6400 West Oakton Street Morton Grove, IL 60053 USA Telephone +1-847-967-2975 or 800-527-2631 http://www.lapmaster.com	Metallographic equipment and consumables
Leco Corporation 3000 Lakeview Avenue Saint Joseph, MI 49085-2396 USA Telephone +1-616-985-5496 or 800-292-6141 http://www.leco.com	Metallographic equipment and consumables Image analyzers Hardness testers Metallographic seminars "Met-Tips" literature
Leica Microsystems AG Ernst-Leitz-Strasse 17-37 35578 Wetzlar Germany Telephone +49-6441-29-0 http://www.leica-microsystems.com	Microscopes, including the Wild, Leitz, Reichert, Jung and Cambridge brands See also: Leica Microsystems Inc. 2345 Waukegan Rd. Bannockburn, IL 60015 USA Telephone +1-847-405-123
LEO Electron Microscopy Ltd. Clifton Road Cambridge CB1 3QH England Telephone +44-1223-414166 http://www.leo-em.co.uk	Scanning electron microscopes (LEO is a joint venture between Leica and Zeiss.) See also: http://www.leo-usa.com
Logitech Ltd. Erskine Ferry Road, Old Kilpatrick Glasgow G 60 5EU, Scotland, UK Telephone +44-0-1389-875444 http://www.logitech.uk.com/	Petrographic equipment and consumables See also: Struers, Inc.—Logitech Product Group

(continued)

(continued)

Company	Product lines
Mager Scientific, Inc. 325 Central St. P.O. Box 160 Dexter, MI 48130 USA Telephone +1-734-426-3885 http://www.magersci.com	Metallographic consumables
Mark V Laboratory, Inc. P.O.Box 310 18 Kripes Rd. East Granby, CT 06026 USA Telephone 888-MARKLAB http://www.markvlab.thomasregister.com	Metallographic equipment and consumables
Materials Engineering Institute (MEI) ASM International 9639 Kinsman Road Materials Park, OH 44073-0002 USA Telephone +1-440-338-5151 or 800-336-5152 http://www.asm-intl.org	Seminars and literature on metallography, ceramics, and related topics in materials science International Metallographic Society (IMS) *Advanced Materials and Processes* journal *Engineered Materials Handbook* series *ASM Handbook* series
Media Cybernetics, LP 8484 Georgia Ave. Silver Spring, MD 20910 USA Telephone +1-301-495-3305 http://www.mediacy.com/	Optimas image analyzers
Metallurgical Supply Company, Inc. 923 FM Road 1959 Houston, TX 77034 USA Telephone +1-281-481-4100 or 800-638-7826 http://www.metsupply.com	Metallographic consumables
Metatech Industries 476, Narayan Peth, Cosmos Bank Building Off Lakshmi Road, Pune-411030 India Telephone +91-20-4489295 http://www.metatechind.com	Metallographic equipment and consumables
National Institute of Standards and Technology United States Department of Commerce Standard Reference Materials Program Bldg. 202, Room 204 Gaithersburg, MD 20899 USA Telephone +1-301-975-6776 http://www.nist.gov	Standard reference materials
Newage Testing Instruments, Inc. 147 James Way Southampton, PA 18966-3817 USA Telephone +1-215-526-2200 or 800-806-3924 http://www.hardnesstesters.com	Hardness testers
Nikon, Inc.—Instruments Group 1300 Walt Whitman Rd. Melville, NY 11747 USA Telephone +1-516-547-8500 http://www.nikonusa.com	Microscopes
Norton Company One New Bond Street P.O.Box 15008 Worcester, MA 01615-0008 USA Telephone +1-508-795-5000 http://www.nortonabrasive.com	Abrasives and saw blades See also: Saint-Gobain-Les Miroirs 92096 La Défense Cedex France http://www.saint-gobain.com
Olympus America, Inc. Two Corporate Center Drive Melville, NY 11747-3157 USA Telephone +1-631-844-5000 http://www.olympusamerica.com	Microscopes See also: Olympus Optical Company, Ltd. Japan http://www.olympus.co.jp/

(continued)

(continued)

Company	Product lines
Pace Technologies, Inc. 200 Larkin Drive Wheeling, IL 60090 USA Telephone +1-888-722-3654 http://www.metallographic.com/	Metallographic equipment and consumables
Ted **Pella,** Inc. P.O.Box 492477 Redding, CA 96049-2477 USA Telephone +1-530-243-2200 or 800-237-3526 http://www.tedpella.com	SEM and microscope accessories
Presi S.A. Tavernolles 38320 Brié et Angonnes France Telephone +33-04-76-7-00-21 http://www.presi.com	Metallographic equipment and consumables
Princeton Gamma-Tech, Inc. C/N 863 Princeton, NJ 08542-0863 USA Telephone +1-609-924-7310 http://www.pgt.com	Image analyzers
Shimadzu Scientific Instruments, Inc. 7102 Riverwood Drive Columbia, MD 21046 USA Telephone +1-410-381-1227 or 800-477-1227 http://www.shimadzu.com	Hardness testers See also: Shimadzu Corporation Kyoto, Japan www.shimadzu.co.jp
South Bay Technology, Inc. 1120 Via Callejon San Clemente, CA 92672 USA Telephone +1-800-728-2233 http://www.southbaytech.com/	Metallographic equipment and electron microscope accessories
Struers, Inc. 810 Sharon Drive Westlake, OH 44145-1598 USA Telephone +1-440-871-0071 or 888-STRUERS http://www.struers.com	Metallographic equipment and consumables Hardness testers *Structure* journal *Metalog Guide* See also: Struers A/S Valhøjs Allé 176 DK-2610 Rodøvre/Copenhagen Denmark Telephone +45-36-70-35-00
TBW Industries, Inc. Forest Grove Rd. Furlong, PA 18925-0336 USA Telephone +1-215-794-8070 http://www.tbw-inc.com	Abrasives
Instron Corporation—**Wilson** Instruments Division 100 Royall St. Canton, MA 02021-1089 USA Telephone +1-781-575-5000 or 800-564-8378 http://www.instron.com/hardness	Hardness testers
Carl **Zeiss,** Inc. One Zeiss Drive Thornwood, NY 10594 USA Telephone +1-914-747-1800 or 800-233-2343 http://www.zeiss.com	Microscopes Hardness testers Image analyzers See also: Carl Zeiss AG Oberkochen, Germany http://www.zeiss.de
Zwick USA LP Kennesaw, GA 30152 USA Telephone +1-770-420-6333 http://www.zwick.com	Hardness testers See also: Zwick GmbH & Co. August-Nagel-Staße 11 D-89070 Ulm Telephone +49-73-05-1 00 http://www.zwick.de

APPENDIX C

Abrasive Size Equivalents

The grit size of an abrasive is equivalent to the quantity of apertures per linear inch (mesh size) in the finest sieve through which the abrasive particles can pass. One linear inch in a sieve includes both the apertures and the wires that form the aperture boundaries, which is why a particle from a 100-grit abrasive is actually somewhat less than 0.01 in. (254 μm) in diameter. Mesh sizes are standardized in ASTM E 11, but the particle size definitions vary somewhat among materials and manufacturers. The Federation of European Producers of Abrasives in France standardizes abrasive particle sizes in many European countries. Particles smaller than approximately 50 μm are considered too small for sieve analysis and are sized by other methods (Ref 1).

Grit/mesh size	Size, µm Diamond	Al_2O_3, SiC, or garnet	B_4C
Coarse grinding			
24	1036	1035	710
30	. . .	930	590
36	712	810	500
46	508	508	350
54	. . .	430	297
60	407	406	250
70	. . .	328	210
80	267	266	177
90	. . .	216	149
100	213	173	125
120	142	142	105
150	122	122	74
Fine grinding			
180	86	70–86	62
220	66	66	53
240	60	54–63	42.5
280	. . .	44	35
320	52	29–32	27.7
400	40	20–23	16.3
600	14	12.5–17	8.3
Polishing			
800	8	9–12	5.5
1000	. . .	5–9	3.7
1200	6	1.5–5	2.5
1500	4	1–1.5	. . .
2000	2	1.0	. . .

REFERENCE

1. L. Dillinger, "Relationship Between Grit Size, Mesh Size, and Micron Size," Met-Tips 15, Leco Corp., 1989

Index

N

O

P

Printed and bound in the UK by
CPI Antony Rowe, Eastbourne

Printed and bound by CPI Group (UK) Ltd, Croydon, CR0 4YY
15/09/2026
14962472-0001